BIOLOGICAL IDENTIFICATION WITH COMPUTERS

BIOLOGICAL DOCUMENTATION AND THE COMPUTER

THE SYSTEMATICS ASSOCIATION
SPECIAL VOLUME No. 7

BIOLOGICAL IDENTIFICATION WITH COMPUTERS

Proceedings of a Meeting held at
King's College, Cambridge
27 and 28 September, 1973

Edited by

R. J. PANKHURST

British Museum (Natural History), London, England

1975

Published for the

SYSTEMATICS ASSOCIATION

by

ACADEMIC PRESS
LONDON · NEW YORK · SAN FRANCISCO

ACADEMIC PRESS INC. (LONDON) LTD.
24/28 Oval Road,
London NW1

United States Edition published by
ACADEMIC PRESS INC.
111 Fifth Avenue
New York, New York 10003

Library of Congress Catalog Card Number: 75–15349
ISBN: 0 12 544850 3

PRINTED IN GREAT BRITAIN BY ROBERT MACLEHOSE AND COMPANY LIMITED
PRINTERS TO THE UNIVERSITY OF GLASGOW

Contributors

AITCHISON, R. R., *Department of Botany, University of Cambridge, Downing Street, Cambridge, CB2 3EA, England.*

BEAMAN, J. H., *Department of Botany and Plant Pathology, Michigan State University, E. Lansing, Michigan 48824, U.S.A.*

GÓMEZ-POMPA, A., *Apartado Postal 70–268, University of Mexico, Mexico 20, D.F.*

GOWER, J. C., *Rothamsted Experimental Station, Harpenden, Hertfordshire AL5 2JQ, England.*

GYLLENBERG, H. G., *The Academy of Finland, Lauttasaarentie 1, 00200 Helsinki 20, Finland.*

HALL, A. V., *Bolus Herbarium, University of Cape Town, Rondebosch C.P., South Africa.*

LAPAGE, S. P., *National Collection of Type Cultures, Central Public Health Laboratory, Colindale Avenue, London NW9 5HT.*

MCNEILL, J., *Research Branch, Plant Research Institute, Canada Department of Agriculture, Ottawa, Canada KIA OC6.*

MORSE, L. E., *Gray Herbarium, Harvard University, 22 Divinity Avenue, Cambridge, Massachusetts 02138, U.S.A.*

MOSS, W. W., *Academy of Natural Sciences, 19th and The Parkway, Philadelphia, Pennsylvania 19103, U.S.A.*

NIEMELÄ, T. K., *Department of Microbiology, University of Helsinki, Helsinki 71, Finland.*

PANKHURST, R. J., *Department of Botany, British Museum (Natural History), Cromwell Road, London SW7 5BD.*

PAYNE, R. W., *Rothamsted Experimental Station, Harpenden, Hertfordshire AL5 2JQ, England.*

ROSS, G. J. S., *Rothamsted Experimental Station, Harpenden, Hertfordshire AL5 2JQ, England.*

RYPKA, E. W., *Lovelace-Bataan Medical Center, 5200–5400 Gibson Boulevard, S.E., Albuquerque, New Mexico 87108, U.S.A.*

SHETLER, S. G., *Smithsonian Institution, Washington, DC 20560, U.S.A.*

WALTERS, S. M., *Botanic Gardens, Cambridge, CB2 1JF, England.*

WILKINSON, C., *Department of Biological Sciences, Portsmouth Polytechnic, King Henry 1 Street, Portsmouth PO1 2DY, England.*

WILLCOX, W. R., *National Collection of Type Cultures, Central Public Health Laboratory, Colindale Avenue, London NW9 5HT.*

Preface

While I was serving on the Council of the Systematics Association in 1971, the suggestion was made that I should organize a meeting to discuss identification of biological specimens by computer. This was held at King's College, Cambridge, on 27th and 28th September, 1973, with about 60 delegates attending. The proceedings of this meeting, presented here, constitute one of the first volumes to be published on this subject.

From the point of view of computer science, our subject comes under what is generally called Pattern Recognition. It must be stated at once, however, that nearly all the techniques which seem useful in this context at present require a description of the object by a human observer. Most of the subjects that biologists want to identify are too complex for automatic description by current methods. This situation could always change in the future. Also, it must be made clear that the final decision of which identification to accept, if any, remains in the hands of the biologist. Machines are not taking over a human role here, but just modelling or mimicking our decision processes. Identification is typically a skill held by just a few individuals, gained after many years of practice. Perhaps the automatic methods will make identifications easier, or simply feasible, for the many for whom identification of specimens is, quite rightly, just a means, and not an end in itself.

The use of computers to help in identification is quite recent, and first becomes discernable in the efforts of several bacteriologists in the early 1960s. The next noticeable development is the appearance of a number of computer programs for constructing diagnostic keys around 1970, and at the present time experiments are being made with a wide variety of different methods. It is interesting that the first impulse to develop numerical methods in classification, as opposed to identification, also came from bacteriologists. The volume of effort in classification by computer far exceeds that put into identification. When one reflects that many biologists carry out identifications daily, and that hardly any complete a biological career and avoid this task, and that the proportion of biologists engaged in classifying things is relatively speaking very small, then this distribution of effort may seem odd. It has been suggested that classification is much more challenging a subject than identification, even if the latter has more practical importance. Readers might like to re-assess this situation after studying these proceedings.

It was decided not to include medical diagnosis within the scope of the meeting. Nonetheless it is interesting to compare the developments in medical diagnosis by computer from the late 1950s onwards with biological identification. Medical work has taken a different course, with much prominence being

vii

given to probabilistic methods. It might be that the problems in the two fields are not as fundamentally different as is often thought. Some references to medical diagnosis are given in the bibliography.

The meeting was planned so that the technical programme was not over-crowded, and included an exhibition and a number of demonstrations of computer programs. Consequently, not every paper which is published here was formally presented at the time. However, the material covered here is all directly ralated to papers presented, discussions, exhibits or demonstrations which took place at the meeting. One paper, which was presented by Dr M. Freudenthal of the National Geological Museum at Leiden, concerning identification applications with a geological data base, was not submitted for publication.

A short film, entitled "Computer Graphics in Fungal Identification", by B. Kendrick, was shown during the meeting. No account of this is given, except for inclusion in the bibliography. I am indebted to Prof. Kendrick for the loan of this film.

My thanks are due to Prof. V. H. Heywood, President of the Systematics Association, for encouragement and assistance throughout, and to Rosemary Aitchison for acting as organizing secretary. We were pleased to welcome Mr J. Gilmour as a session chairman. A grant made by the Royal Society towards speakers' travel expenses is gratefully acknowledged.

R. J. P.

May, 1975

Contents

Historical Introduction

Historical Introduction

1 | Traditional Methods of Biological Identification

S. M. WALTERS

University Botanic Garden, Cambridge, England

Abstract: Identification of biological specimens can be defined as the practice of assigning the specimens to known, named taxa. It is a necessary activity which most biologists undertake at some time, but in which taxonomists are specially concerned since they produce both the classifications and the tools for identification. Some of these tools, especially the artificial dichotomous key, have achieved particular prominence, but surprisingly little attention has been paid to the relative merits of different tools. The advent of computers and numerical taxonomy have been beneficial in stimulating taxonomists to ask some of these practical questions. The paper is illustrated by reference to the classification of the Umbelliferae.

Key Words and Phrases: identification, dichotomous keys, Umbelliferae, history of taxonomy, taxonomic description

The practice of identification is necessarily such a common experience for all biologists who are working with whole organisms that it seems very appropriate to begin a meeting devoted to automatic identification with a brief outline of the methods actually employed and the history of their use. My only reluctance to do this arises from the way in which my own knowledge of the subject is restricted to the higher plants, as I am aware that we have at this meeting specialists in several other groups. I believe, however, that the important aspects of the traditional method can be conveniently illustrated from the history of botanical classification, and that much of my thesis could have been similarly illustrated by zoological examples.

Identification of biological specimens can be defined as the practice of assigning a given specimen to a known, named taxon. All biologists are involved in identification (if only as laymen in the daily round of affairs), but the taxonomist is specially concerned, since he produces or alters the classifications of organisms as well as providing the tools by which his fellow-scientists can

Systematics Association Special Volume No. 7, "Biological Identification with Computers", edited by R. J. Pankhurst, 1975, pp. 3–8. Academic Press, London and New York.

identify their specimens. It is very instructive to look at the history of taxonomy, and to attempt to trace the interrelations of naming, classification, and identification.

Logically, it would seem that this order of activity must be operating: a taxon is recognized by naming, its position is decided in a hierarchical classification, and then specimens can be assigned to it by a procedure of identification. The history of biological taxonomy does not, however, reveal this process in such a logical sequence, and a little thought soon tells us why it cannot be so. The most obvious complicating factor in the process is that taxonomic knowledge is increasing all the time, so that attempts at identification continually show inadequacies in the existing system of names and hierarchical groups, and new taxa and systems are made to accommodate the new knowledge. For those who are interested in pursuing implications of these thoughts, I unhesitatingly recommend an excellent paper by E. G. Voss (1952) and the many references given in it.

To illuminate our subject, I have selected a single, very familiar group of flowering plants, the members of the carrot family Umbelliferae. There are several reasons for my choice; but a very special reason is that we have just celebrated the tercentenary of Robert Morison's monograph on this family, published in Oxford in 1672, and conveniently described by Hedge (1973).

Morison's monograph is an impressive work which reminds us that, in cases where a modern flowering plant family has many common European representatives, the naming and classification of the family took shape in Medieval Europe, long before Linnaeus and the eighteenth century standardization of nomenclature and classification. The implications of this "European bias" for Angiosperm taxonomy in general I have discussed elsewhere (Walters 1961, 1962), and these general themes lie outside the present field of discussion. The Umbelliferae, however, are not only common, but have for centuries been known for their culinary, medicinal and even poisonous properties; for these reasons also they were the subject of description and illustration in Classical and Medieval writings (Fig. 1), and their correct identification was a matter of some practical importance (see French, 1971). It is, therefore, not surprising to find that Morison provided detailed illustrations of the "seeds" (actually the fruits) of many different kinds of umbellifers (Fig. 2). What is perhaps less expected is that the monograph also includes bracketed diagrams which function to some extent both as a classificatory device ("conspectus") and as an identification tool or "key" (Fig. 3).

According to Voss, several biological writers in the second half of the seventeenth century used such diagrams, and Nehemiah Grew described their use for identification as early as 1676. Curiously enough, the term "clavis" (key) was apparently not used in connexion with such diagrams until Linnaeus so used it in 1736 (and then with reference to a diagram in which he was classifying *botanists*, not plants!). The credit for explicit and systematic use of modern artificial dichotomous keys *for identification* is usually given to Lamarck in his "Flore Française" (1778), and after this pioneer work most nineteenth century Floras supplied such keys as a matter of course.

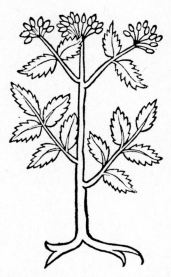

Fig. 1. "Carvi" (Caraway), illustration from Herbal "Ortus Sanitatis", Mainz 1491 (taken from Arber (1938) p. 165).

It would therefore appear from a comparative study of botanical writings over the seventeenth, eighteenth and nineteenth centuries that the *description* and the *illustration* were the earliest identificatory aids, and that the modern, standard, artificial key gradually developed from a diagram which, by grouping and differentiating the different "kinds" of plants (in the case of Umbelliferae most of these "kinds" correspond to the modern genera), served the purposes both of classification and identification. The rigid, logical separation between a synopsis of classification or *conspectus* on the one hand and an artificial *key* on the other seems to have been relatively late in developing. Indeed, examples could still be found in recent Floras where the "keys" provided seem to be uncomfortably attempting to satisfy both these requirements and achieving neither aim as a result.

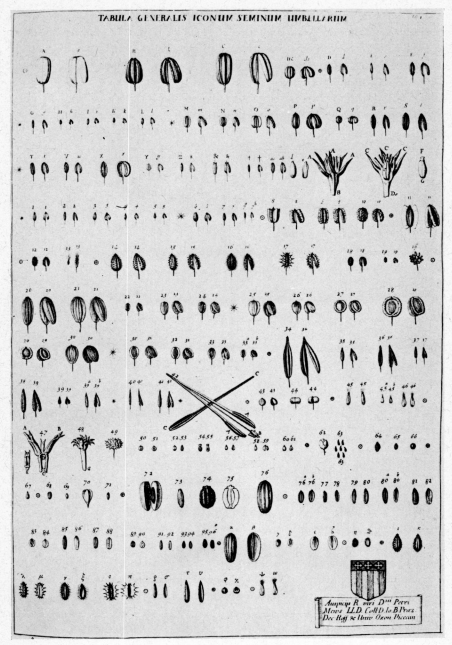

FIG. 2. Fruits of Umbelliferae, from Morison (1672).

To conclude my survey, I can turn to the treatment of the Umbelliferae in volume 4 of the "Flora of Turkey" (Davis *et al.*, 1972), published exactly three centuries after Morison. The first thing to say is that the continuity (some would

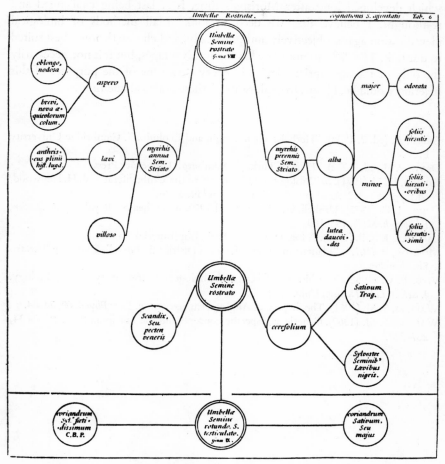

Fig. 3. Diagram of Umbelliferae (Part), from Morison (1672).

say conservatism) of botanical taxonomy is such that Robert Morison, if he were to come alive again, could use this book without much difficulty. Many of the generic names are the same; the nomenclature remains in Latin; there are detailed illustrations of the fruits, and there are "diagrams" or keys to aid identification. (The text is in English, which should also cause him little difficulty!) The "Flora of Turkey" authors have, however, made one significant

8   *S. M. Walters*

departure: they have produced a "multi-access key" which enables a much freer use of characters for generic identification. I feel sure Morison would have approved; it would seem that the rigid orthodoxy of the dichotomous key, which developed long after Morison's time, is at last being challenged and re-thought. I believe that we need to look at all methods of biological identification again, objectively and practically, and choose the ones best suited to our task. The dichotomous key has many advantages, but it is not necessarily the only or the appropriate device in every case. One of the purposes of this meeting would be, I hope, to explore afresh these practical questions.

REFERENCES

ARBER, A. (ed. 2, 1938). "Herbals: Their Origin and Evolution". Cambridge University Press.

DAVIS, P. H., ed. (1972). "Flora of Turkey". Edinburgh University Press.

FRENCH, D. H. (1971). *In* "Biology and Chemistry of the Umbelliferae" (V. H. Heywood, ed.), pp. 385–412. Academic Press, London and New York.

HEDGE, I. C. (1973). Umbelliferae in 1672 and 1972. *Notes from the Royal Botanic Garden, Edinburgh.* **32**(2), 151–160.

LAMARCK, J. B. P. (1778). "Flore Française". Paris, Imp(rimerie) Royale.

MORISON, R. (1672). "Plantarum Umbelliferarum Distributio Nova". Oxonii, e Theatro Sheldoniano.

VOSS, E. G. (1952). The history of keys and phylogenetic trees in systematic biology. *J. Scient. Labs. Denison Univ.* **43**, 1–25.

WALTERS, S. M. (1961). The shaping of Angiosperm taxonomy. *New Phytol.* **60**, 74–84.

WALTERS, S. M. (1962). Generic and specific concepts and the European flora. *Preslia* **34**, 207–226.

Survey

2 | Recent Advances in the Theory and Practice of Biological Specimen Identification

LARRY E. MORSE

Department of Biology, Harvard University, Cambridge, Massachusetts, U.S.A.

Abstract: This review considers modern methods of specimen identification in systematic biology and natural history, and discusses their general theoretical basis and their use (frequently with computer support) in specific applications. Computer-based procedures for specimen identification often employ *polythetic* methods in which no single characteristic of the unknown specimen is considered sufficient evidence for excluding some taxon from the set of possible identifications. Many of the new methods are also *polyclave* procedures, offering the user a choice of characters for each instance of identification. Computers are readily used to help construct and edit traditional dichotomous keys, but computer-stored keys offer few advantages over printed ones. Simultaneous character-set methods have been developed for cases such as microbiological identification where it is preferable first to make observations of a number of characters together, then to use these observations all at once. Possibilities for fully automated identification have been explored to some extent, but such methods are not yet used in any routine work. The conclusion is reached that many important new methods of specimen identification are now ready for widespread use, but our present taxonomic knowledge is in most cases inadequate or too poorly organized to support them.

Key Words and Phrases: specimen identification, information systems, dichotomous keys, decision trees, character weighting, discriminant analysis, multiple-entry keys, polyclaves, punched-card keys, pattern recognition, data banks, taxonomic knowledge

INTRODUCTION

A few years ago, biological specimen identification was a topic generally untouched by the computer revolution. Now, as clearly shown by this Systematics Association symposium, the study of computer-assisted specimen identification in biology is a diversified and maturing discipline, beginning to reach practical, everyday application as well as theoretical sophistication. The rapid development of this field is particularly evident when we compare the

---

Systematics Association Special Volume No. 7, "Biological Identification with Computers", edited by R. J. Pankhurst, 1975, pp. 11–52. Academic Press, London and New York.

diversity of papers in the present volume with the consideration of specimen identification in earlier symposia on computer techniques in systematic biology (e.g. Gómez-Pompa and Squires, 1969; Cutbill, 1971a). In this respect, we can also contrast the detailed treatment of specimen identification in Sneath and Sokal's recent (1973) numerical taxonomy book with the speculative discussion in their earlier one (Sokal and Sneath, 1963).

In this review, I discuss and compare a number of modern approaches to specimen identification; most of these involve computer implementations. I also briefly evaluate the appropriateness of these methods for various applications, and make suggestions for future work, particularly comparative studies. I here avoid detailed consideration or evaluation of specific methods and algorithms, and also avoid formal theory; Sneath and Sokal (1973, pp. 381–408) provide a thorough consideration of the theoretical basis for these diverse modern procedures of specimen identification. This survey draws heavily upon my earlier work in this area at Michigan State University (Morse, 1974a), and also benefits from extensive discussions and correspondence of the past several years with various others interested in modern methods of specimen identification, particularly J. H. Beaman, A. V. Hall, R. J. Pankhurst, S. G. Shetler, P. H. A. Sneath, and the late J. A. Peters. I thank several participants in the present symposium for advance copies of their papers.

The four outstanding papers from the Jerusalem symposium on computer-aided determinative bacteriology appeared too recently for discussion in the text of this review, but will be mentioned briefly here. Hill (1974) provides a detailed discussion of various theoretical aspects of identification; his tabular comparison of several measures of character importance will be particularly useful. Lapage (1974) discusses his continuing experiences with the everyday use of probabilistic identification in clinical bacteriology. Sneath (1974) presents a thorough consideration of the importance of test reproducibility in identification, including both theoretical aspects and experimental results. Krichevsky and Norton (1974) discuss data-handling aspects of computer-assisted identification, particularly problems of coding and standardization.

MAJOR METHODS OF SPECIMEN IDENTIFICATION

Table I presents a comparison of the several approaches to biological specimen identification which will be discussed in this paper. Omitted here are non-algorithmic techniques such as immediate recognition or referral of a specimen to an expert; these are beyond the scope of our survey. Also omitted are methods for placing taxa (rather than specimens) into higher groups (e.g. Lawrence and Bossert, 1967; Machol and Singer, 1971; Kamp, 1973; McHenry,

TABLE I. Comparison of major methods of specimen identification

	Computer-stored keys	Computer-constructed keys	Simultaneous character-set methods	Polyclaves and other sequential methods	Fully automated identification
Human observer required for each specimen?	yes	yes	yes	yes	no
Taxon/character data matrix required?	no	yes	yes	yes	sometimes
Large number of characters per specimen generally requested?	no	no	yes	no	(observed automatically)]
Specimen generally identified as wrong taxon if correct taxon not initially considered?	yes	yes	usually not	usually not	sometimes
Procedure easily adapted to polythetic mode?	no	no	yes	yes	yes
Procedure particularly useful for hybrids or overlapping taxa?	no	no	yes	yes	sometimes
User free to select characters he wants for each specimen?	no	no	no	yes	(characters selected automatically)
Consideration of relative character conveniences possible?	yes	yes	no	yes	yes
Consideration of relative taxon abundances possible?	yes	yes	yes	yes	yes
Can procedure be used effectively with batch-processing computers?	no	yes	yes	no	yes
Is procedure adaptable to easy use in the field without a computer?	no	yes	yes	yes	no

1973), and methods for reassigning individuals in attempting improvement of a given classification (e.g. Demirmen, 1969). Leenhouts (1966) and Morse (1971) provide more thorough discussion of noncomputerized identification methods than given here. Pankhurst (1974) has recently presented a comparison of computer-based identification procedures. He considers three basic approaches: keys and polyclaves, matching methods, and methods based on Bayes' theorem. Here I consider the Bayesian methods as advanced interactive polyclaves, and on the other hand discuss dichotomous keys and multiple-entry polyclaves separately. Pankhurst's "matching methods" correspond to my "character-set" methods.

The important distinction between *monothetic* and *polythetic* procedures deserves discussion here, since nearly all methods of specimen identification may be used in either mode. In monothetic identification, a single character or a particular, pre-specified combination of characters is both a necessary and a sufficient criterion for assigning a specimen to a particular taxon. Most dichotomous keys are monothetic, since in them there is but one route to each taxon in the key; a specimen must show all of the necessary characters in the proper states to be keyed out as a particular taxon. On the other hand, the measures of "overall similarity" promoted by numerical taxonomists do not depend on any particular characters; identification methods based on equivalent principles are polythetic.

The distinction between monothetic and polythetic methods may be clearer when criteria for excluding a specimen from a given taxon are considered. In monothetic identification, a particular single character ("key character"), or the lack of such a character, can be sufficient to exclude a specimen from some taxon, regardless of how well the specimen agrees with members of the taxon in other ways. However, in polythetic identification, the specimen could still be identified with that taxon, providing it shows sufficient "overall similarity", despite the difference in the one "key character" of the group. For further discussion of the polythetic taxon concept and its implications for specimen identification, see Mayr (1969, pp. 82–83) or Sneath and Sokal (1973, pp. 20–23). Zadeh's work on "fuzzy sets" is also of interest here as a quantification of some of these concepts (Zadeh, 1965; Bellman and Zadeh, 1970); Flake and Turner (1968) and Bedzek (1974) discuss his work in a taxonomic context.

COMPUTER-STORED KEYS

Systematists first seeing an interactive computer dialog often think of using the computer to step through a traditional dichotomous key, with the computer asking a question, awaiting an answer, then asking the question appropriate for that response, and so forth. The computer programming for such a key is trivial, primarily involving a PRINT–INPUT–IF sequence. Bossert's (1969) implementation of a computer-stored key to Polynesian ants illustrates this approach, as does Kendrick's (1972) use of interactive computer graphics in stepping through an illustrated key to a group of fungi.

Although computer-stored keys are a good test of one's computer-programming ability, in their simplest form they offer no advantages over printed keys, and have the major disadvantage of requiring access to and use of a computer for each identification. Hence, until computers become cheaper than books, computer-stored keys can be justified only when additional capabilities such as error correction (cf. Sokal and Sneath, 1966) are offered. Generally, a computer-based interactive polyclave procedure (discussed below) should be preferable to a computer-stored key.

A topic related to computer-stored keys is computer-assisted key editing, explored a few years ago by Morse *et al.* (1968, 1969). For this, the text of a dichotomous key is recorded on computer cards in a way which facilitates revision and correction. A computer program then prints the corrected key in traditional indented or bracketed form. Further work could link the key-editing program to an on-line text editor (cf. Van Dam and Rice, 1971) to provide on-line revision of keys.

COMPUTER–CONSTRUCTED KEYS

Traditional dichotomous identification keys can be readily constructed by computer from machine-readable files of taxonomic information. In fact, in some ways the challenge and difficulty of this topic lie more in the data collection and coding than in the computer algorithm itself. Character definition and coding are particularly troublesome areas here, as in taxonomic information processing generally (Jičín, 1972; Morse, 1974b). Nevertheless, substantial progress has been made in recent years in the construction of keys by computer, as is clear from the present volume. Not long ago, I provided a longer discussion of this topic (Morse, 1971), and Pankhurst (1974) has made a more recent survey, including a tabular comparison of four key-constructing programs. The long history of the idea of the dichotomous identification key was reviewed by Voss (1952), and common difficulties in the use of keys were discussed by Harrington and Durrell (1957) and by Voss (1972, pp. 30–35), amongst others. Sneath (1957) concluded that monothetic dichotomous keys are inappropriate for identification in microbiology.

The commonest approach to key construction by computer is a recursive algorithm which repeatedly divides sets of taxa into pairs of mutually exclusive subsets on the basis of one or more taxonomic characters. Each application of the partitioning procedure produces one dichotomy or couplet of the key. Generally, a character is selected at each dichotomy which divides the relevant taxa as equally as possible between the two subgroups being formed. Osborne's (1963b) consideration of the theory of keys supports this equal-division procedure, which can also be developed from information-theory arguments (cf. Maccacaro, 1958; Möller, 1962; Jičín, 1974b). Nievergelt (1974) surveys key-like techniques (binary search trees) in an information–retrieval context.

The actual selection of a dichotomizing character at some point in constructing a key can be readily made by calculating and optimizing some kind of dichotomizing score for each of the available characters when considered over the set of relevant taxa. (Actually, the score should be calculated for each pair of contrasting states within each of the characters, when the characters have more than two states.) It is in the details of this character-choice calculation that the available key-constructing programs differ most significantly, although for all the general results they are the same. For example, an extension of Gyllenberg's (1963) "separation figure" (S) can be used to rank characters for use in a key. In the simplest case, namely binary (two-state) characters having no variable or missing values, the S value is simply the product of the number of taxa showing each of the states of a character. For example, if we have 50 taxa, and they divide 25–25 for one character but 30–20 for another, the respective S scores for the two characters are 25×25 (625) and 30×20 (600). Maximizing the Gyllenberg S optimizes the evenness of a dichotomy. Niemelä *et al.* (1968) presented a key-constructing program for this simplest binary case, and equivalent algorithms have been developed in non-biological areas (e.g. Ledley and Lusted, 1959b). For cases involving variable or missing values, and multistate or quantitative characters, more complicated character-evaluation formulas are needed (Morse, 1971, 1974a). Several such formulas have been proposed in recent years, including those of Hill and Silvestri (1962), Sneath (1962), Gyllenberg (1964), Dybowski and Franklin (1968), Lapage and Bascomb (1968), Jičín *et al.* (1969), Hall (1970), Pankhurst (1970a), Barnett and Pankhurst (1974), Dallwitz (1974), Gower and Barnett (1971), and Barnett (1971). A detailed comparison of these methods, considering both theoretical aspects and costs in actual use, would be quite helpful. Where these tests are equivalent, then the fastest should be used; where they are different, they could be offered as user options in future key-constructing programs, much as Peters (1968) did for alternative similarity measures in his biogeography program. Further work on methods for comparing quantitative characters is needed; the present tests have little strong theoretical basis. Further study is also needed of the question of character-convenience weighting and taxon-frequency weighting, and the effect of these considerations on the practical efficiency of identification keys (cf. Morse, 1971; Dallwitz, 1974). Finally, additional investigation can be done on what it is that makes a particular character "good" at some point in a key. This may have no simple answer, since a number of subjective factors are involved (cf. Metcalf, 1954, or Mayr, 1969, pp. 276–281), as well as objective, mathematical considerations.

Key construction by computer is currently being studied primarily by A. V. Hall, R. J. Pankhurst, and myself, and our work to date is described in a number of papers.* Hall and I intend our programs only to be aids to a systematist writing a key to a group of taxa; the computer is used basically to suggest different ways in which a key to the group might be made. Experimentation with weightings for character conveniences, taxon frequencies, and other parameters is particularly helpful here in obtaining several rather different keys from the same basic data. (In an interactive key-constructing program, some of these parameters might even be changed on-line as the key is produced. A fully interactive key-constructing program could also be produced, in which the computer would suggest several characters for the next dichotomy but leave it to the systematist to pick one.) After obtaining several computer-produced keys, the researcher can compare them and select the best ones for his purposes, then improve these by rephrasing the leads and adding supplementary characters. This can be done either by hand or with a key-editing program. Figure 1 shows a computer-constructed key, ready for further editing.

Pankhurst has tried taking automated key construction one step further, in aiming to obtain publication-quality keys directly from the computer (e.g. Barnett and Pankhurst, 1974). In his program, character phrasing and supplementary-character inclusion are done automatically, and the resulting keys under some circumstances can be incorporated directly into a taxonomic work. However, Watson and Milne (1972) found it necessary to edit keys produced by Pankhurst's program prior to publication, whereas Webster (1969, 1970) published them directly.

SIMULTANEOUS CHARACTER-SET METHODS

The techniques discussed in this section offer a diversity of approaches to biological specimen identification. Some of them have been explored reasonably well in biological applications; others are better developed in different areas of

---

* Hall (1970, 1973 and this volume, p. 55); Pankhurst (1970a, b, 1971); Barnett and Pankhurst (1974); Pankhurst and Walters (1971); Saila (1970); Morse (1968, 1969, 1971, 1974a); Beaman (this volume, p. 267); Furlow and Beaman (1971); Morse *et al.* (1968, 1969); Solbrig (1970, pp. 200–201); Shetler *et al.* (1971).

USED   5.00 UNITS
RUN

VKEY   17:36   S2 WED 08/27/69

IN DPRINT

MICHIGAN ØAKS MATRIX -- MARY BETH MØØRE
USE "MICHIGAN ØAK CHARACTERS" MAY 20, 1969

1. LØBES ØR TEETH ØF LEAVES BLUNT (SØMETIMES ACUTE);
   NØT BRISTLE-TIPPED
   2. LEAVES DISTINCTLY LØBED
      3. UPPER CUP-SCALES LØØSE, FØRMING A CØNSPICUØUS
         FRINGE ØR LØNG-AWNED
            .....Q. MACRØCARPA
      3. UPPER CUP-SCALES TIGHTLY APPRESSED, NØT FØRMING A
         DEFINITE FRINGE ØR AWNED
            .....Q. ALBA
   2. LEAVES CØARSELY TØØTHED, WITH 3-14 TEETH ØN EACH
      SIDE, THE SINUSES EXTENDING LESS THAN ØNE-THIRD
      ØF THE WAY TØ THE MIDVEIN
      4. LEAVES WITH ACUTELY PØINTED TEETH
         5. LEAVES WITH 3-7 TEETH ØN EACH SIDE
            .....Q. PRINØIDES
         5. LEAVES WITH 8-13 TEETH ØN EACH SIDE
            .....Q. MUEHLENBERGII
      4. LEAVES WITH RØUNDED ØR CRENATE TEETH
            .....Q. BICØLØR
1. LEAVES ØR THEIR SHARP LØBES ØR TEETH
   BRISTLE-TIPPED
   6. NUT ØF ACØRN ØVØID TØ ELLIPSØIDAL
      7. TERMINAL BUDS DISTINCTLY ANGLED IN CRØSS-SECTIØN
            .....Q. VELUTINA
      7. TERMINAL BUDS CIRCULAR ØR ØNLY SLIGHTLY ANGLED
         IN CRØSS-SECTIØN
         8. LØNGEST LØBES ØF LEAF SHØRTER THAN, TØ ABØUT
            EQUALLING (NEVER TWICE AS LØNG AS) THE BREADTH ØF
            THE BRØADEST MEDIAN PØRTIØN ØF THE BLADE
               .....Q. RUBRA
         8. LØNGEST LØBES ØF LEAF TWØ TØ SIX TIMES AS LØNG AS
            BREADTH ØF NARRØWER PART ØF MEDIAN PØRTIØN
            ØF BLADE
               .....Q. ELLIPSØIDALIS
   6. NUT ØF ACØRN GLØBØSE ØR HEMISPHERICAL
      9. LEAVES ENTIRE
            .....Q. IMBRICARIA
      9. LEAVES NØT ENTIRE
         10. ACØRN-CUP SHALLØWLY TØ DEEPLY SAUCER-SHAPED,
             CØVERING ØNE-FØURTH TØ ØNE-THIRD THE NUT
               .....Q. PALUSTRIS
         10. ACØRN-CUP TURBINATE ØR HEMISPHERIC, CØVERING AT
             LEAST ØNE-THIRD TØ ØNE-HALF ØF THE NUT
               .....Q. CØCCINEA

NØRMAL ENDING ØF DICHØTØMØUS KEY

FIG. 1. Key to species of oak (*Quercus*) in Michigan, constructed and printed by computer using program by Morse (1969) and data prepared by an undergraduate biology student. In producing this key, all character couplets were given equal character-convenience weightings, so leaf, fruit, and twig characters are used indiscriminately. (Since these data did not have expert review, the key should not be regarded as taxonomically definitive.)

decision analysis, such as medical diagnosis, psychology, or pattern recognition. *Character-set methods* are those in which values or states of a predefined set of characters are recorded for the unknown specimen, then submitted simultaneously to a computer program (or equivalent) for analysis. Then, one or more possible identifications are suggested, or the conclusion might be reached that the specimen appears to differ from all the taxa being considered as possibilities.

These character-set methods may be monothetic or polythetic, and qualitative or quantitative. Some of the quantitative character-set methods, such as discriminant analysis, are among the best tools available for identifications involving apparently overlapping (fully polythetic) taxa. However, all these methods have the disadvantage of requiring the observer to record a fairly large number of characteristics of each specimen, generally considerably more characters than would be needed to identify the specimen with a sequential method such as a polythetic polyclave.

There are three broad areas of systematic biology where the simultaneous character-set methods of identification have proven particularly important. These are identification in microbiology, identification involving possible hybrids, and possibilities for fully automated identification. The first two of these will be discussed next; fully automated identification is considered in a separate section.

In microbiology, the characters generally used for identification are biochemical and physiological tests made with living cultures of the unknown isolate; for further discussion see Steel (1962) or Sneath (1969). These tests take hours or days to perform, but several can be done in parallel on separate cultures of the unknown. Thus, for prompt results it is preferable to perform a number of tests in parallel, then attempt to identify the organism once the results of all these tests are available. To identify bacteria with a sequential method such as a dichotomous key could take weeks, if the tests were performed one by one as called for in the key. For microbial identification, and especially for urgent clinical work, various character-set methods are now widely used. These include published tabular keys (e.g. Cowan and Steel, 1965), various mechanical devices (e.g. Cowan and Steel, 1960; Olds, 1970), and a number of computer-based procedures. Also, methods for optimization of the character set for a given application are being developed.

For different reasons, the identification of hybrids is also difficult with ordinary keys and polyclaves. Frequently, hybrids are simply ignored by the author of a key. However, since hybrids (especially plant hybrids) are a common part of the natural biota, and are frequently encountered in agriculture and horticulture, they cannot be ignored when developing comprehensive identification

20   L. E. Morse

systems. First-generation hybrids may sometimes be so distinctive that for identification purposes they may be treated simply as additional, separate taxa (cf. Wagner, 1969). More often, individuals of possible hybrid origin are readily placed to species group by ordinary means, but their placement within a small group of similar species is much more difficult. The occurrence of backcrosses, hybrid swarms, and introgression further complicate the scene, for these phenomena can produce a spectrum of individual variation ranging from one parental type to the other, or even beyond; Rollins and Solbrig (1973) discuss a typical case. For identification in such two-species cases, several quantitative character-set methods have been developed in recent decades, ranging in complexity from the straightforward scatter diagrams and hybrid indices of Edgar Anderson (1949, pp. 81–101) to discriminant analysis and other computer-based techniques discussed below. Anderson's simple, pencil-and-paper methods should not be ignored just because they look easy; Goodman (1967) concluded they were just as effective as discriminant functions in the case he studied. For selecting among several possible parentages of a suspected first-generation hybrid, Anderson's method of extrapolated correlates may be helpful; Tucker (1968) gives a detailed example of such a case. However, when complex multiple-species crosses are involved, perhaps in a situation such as the one hypothesized by Camp (1945) for North American blueberries (*Vaccinium*) or by Wagner (1954) for Appalachian *Aspleniums*, powerful quantitative techniques as well as an experienced systematist may be needed to place individuals in the overall picture.

A natural adjunct to character-set identification methods is character-set optimization, i.e. selection of the optimum subset of characters which still provides sufficient information for identification of specimens of any of the taxa being considered as possibilities. The simplicity of the problem is deceptive, especially in practical biological applications (cf. Newell, 1970; Piguet and Roberge, 1970). The major difficulty is that the mathematical minimal set of characters may not be the most desirable set on biological grounds, for the ease of observation and the reliability of the characters must also be considered. Also, it is helpful to find a set of characters that can all be observed by similar procedures, such as character sets containing only biochemical tests or only features observable by scanning electron microscopy. Alternative choices of states or ranges within characters can also be considered in forming character sets (cf. Bisby, 1970; Hall, 1970; Hicks, 1973); finer splitting of states significantly improves the power of a character, but also decreases the reliability with which observations can be associated with the proper character state. There is an extensive literature on character-set optimization in biological contexts (e.g.

Gyllenberg, 1963; Rypka *et al.*, 1967; Niemelä *et al.*, 1968; Willcox and Lapage, 1972), but no review can be attempted here. Despite the attention to character-set optimization, the problem apparently remains unsolved in any general sense.

Monothetic Character-set Matching

The simplest character-set method of identification to visualize is an exact-match (monothetic) tabular search, in which a taxon/character data table is searched taxon by taxon for an exact match with the list of characters of an unknown specimen. Actually, either a specimen/character data matrix or a taxon/character one can be used; taxon/character matrices are preferable since character combinations can be included that are not represented by any single specimen in the reference set (Morse, 1971). Printed taxonomic tables are included in many reference works in systematics and natural history, and are readily used monothetically for specimen identification. Osborne (1963a) and Baer and Washington (1972) have proposed methods for organizing printed tables to facilitate search without a computer, and Rypka's (1971) truth tables can also be readily used without computer support. One example of a computer-based monothetic (exact-match) tabular search method in biology is the computer-stored key to Latin American snake genera developed by Peters (1969a, b; Peters and Collette, 1968). Although the program is internally a computer-stored dichotomous key, it requests a prespecified set of 14 characters from the user for every specimen and is therefore in practice a character-set method. (Peters later developed a computer-based polyclave for the same group, allowing free character choice by the user.) Two other examples of exact-match tabular searches are the program developed by Corlett *et al.* (1965) for bacteriological data, and the simple key to selected microorganisms developed for teaching purposes by Gasser and Gehrt (1971).

Polythetic Character-set Methods

In biology, polythetic character-set methods have been explored and developed much more frequently than monothetic ones, and several powerful techniques are now available. Polythetic character-set procedures can be thought of as those in which similarities rather than exact matches are determined over a specified set of characteristics. A simple but representative approach, explored in detail by Pankhurst elsewhere in this volume (p. 79), is to rank the taxa by the number of character-state agreements between their descriptions and the unknown specimen, considered over a predetermined set of characters. Taxa having this similarity score highest are suggested as likely identifications. Pankhurst's program prints a proportion of the original list at this stage; he suggests the identification be sought among these by traditional means.

An approach similar to Pankhurst's was taken by Walker *et al.* (1968) for pollen identification with a morphological data bank. They provide the capability of allowing for missing information, always a problem when using a dichotomous key or a character-set method with fossils or other poorly preserved materials. In their approach, Walker's group calculates a similarity score based on matches within groups of characters rather than matches on individual characters. Possibilities having at least one match in each of their eleven character categories are printed first, followed by those with matches in ten of the eleven groups, and so forth. Thus, differences between the specimen and some taxon for all characters within one of the eleven groups excludes that taxon from the set of taxa most similar to the specimen, regardless of how similar they may be for characters in other groups.

Hall (1969) discusses specimen identification with a more complicated matching program. His method includes the useful technique of discontinuing a particular comparison once a preset amount of disagreement is exceeded. However, his paper does not explain how those limits are determined for particular cases.

Gyllenberg (1964, 1965; Gyllenberg and Niemelä, this volume) has developed a polythetic character-set method using geometric models. In the Gyllenberg method, a multidimensional "identification space" is established with one axis for each character in the character set being employed. Each taxon being considered as a possible identification is assigned a hyperspherical region in this identification space, with a center and radius determined by computer program from the taxon's description. In general, two taxa may be distinct or may overlap, but should not occupy identical regions. Actually, two radii are calculated for each taxon; the smaller one determines the taxon's "identification region", and a slightly larger one determines its "neighborhood". To make an identification, a point for the unknown specimen is located in the identification space on the basis of the specimen's characteristics. The distance from that point to the centers of the various taxon clusters is then calculated, and the relation of the unknown to the two nearest clusters is examined. If the unknown lies within the identification region of the closest taxon, but is outside the neighborhood and identification regions of the next-closest taxon, then that closest taxon is suggested as an identification. If the unknown lies within the neighborhood regions of both the closest taxa, the specimen is judged an intermediate. For specimens that are neither identified nor called intermediates, two possibilities are considered. If the unknown is within the neighborhood region of one taxon but not the other, it is called a "neighbor" of that taxon. However, if the unknown is outside the neighborhood regions of both the closest taxa (and hence outside the neighborhood regions of all the taxa being considered), it is called an "outlier" and probably represents a taxon not included in the identification scheme. A number of procedures may be used to calculate the radii for Gyllenberg's scheme, including Euclidian distances, probabilities, and correlation coefficients, as Gyllenberg and Niemelä discuss in their paper (this volume). Quantitative studies of taxon circumscriptions (e.g. Gilmartin, 1969, 1974) should provide evidence on the appropriate sizes for these radii in various cases. While Gyllenberg's method uses hyperspheres, other "shapes" for taxa might prove more appropriate.

A powerful method for relatively small character sets, the truth table, has been applied to microbial identification by Rypka (1971 and this volume; Rypka and Babb, 1970). In the truth-table method, a small set of binary or multistate characters is specified, and a sequential list of all possible combinations of character states for these characters is constructed. For each character-state combination logically possible, the appropriate identification (if any) is listed. For example, five two-state (binary) characters give a truth table having 2^5, or 32, entries, while one for ten binary characters has 2^{10}, or 1024, entries. Clearly, truth tables are only practical for small character sets. Another limitation of the truth table approach is that combinations of states within a particular character are not permitted as separate entries, since allowance for all multiple combinations of character states would expand the table intolerably. Hence, any variability within the specimen to be identified can be handled only by considering each combination of the variable characters' states to be a separate unknown; if more than one taxon is obtained as a possible identification by this procedure, the truth-table method cannot distinguish among them. In general, truth tables are polythetic, since a taxon can have several entries with no single character in common. Small truth tables can be printed and used manually, but larger ones require computer support for convenient use. Being a table look-up procedure rather than a tabular search, truth tables can be extremely efficient computationally compared with other polythetic character-set methods when a large number of taxa are involved. However, the severe size constraints of truth tables restrict their use to small character sets.

Several probabilistic character-set methods of identification have been developed, and the various sequential probabilistic methods discussed later can also be used in a character-set mode if desired. A straightforward probabilistic technique for identification using sets of quantitative characters is described by Davidson and Dunn (1966), and an example of its use is discussed in detail by Stephenson (1971). In this method, several quantitative characteristics are selected, and data from a number of specimens of each taxon are collected. Each character is next segmented into a series of small ranges of values. Then, for each taxon and each character, statistical calculations are made of the probability that a specimen of that taxon will have the measurement for that character in each of the various segments of that character's range. Once all these probabilities have been calculated, identifications of unknown specimens can be attempted by examining the appropriate characters of the unknown, then multiplying the various tabulated probabilities to give a "combined probability" score for each of the taxa being considered. Statistical tests are then used to determine if one of the taxa is a likely identification of the unknown specimen. Vogt and McPherson (1972) describe an approach similar to Davidson and Dunn's, but for qualitative data. Various qualitative polyclave procedures can also be used for character-set identifications using qualitative data.

Discriminant analysis is a powerful character-set technique that has been widely used in biology since its development by R. A. Fisher (1936) for a challenging problem involving three *Iris* species, posed by Edgar Anderson. However, discriminant analysis demands quantitative data from numerous specimens, and requires heavy calculations, so it should not be undertaken without good reason. Sneath and Sokal (1973, pp. 400–408) gave a detailed discussion of discriminant analysis and related methods, and Sokal (1965) and Crovello (1970) cited many specific examples of their use. Goodman (1967) tested several discriminant methods with hybrids of known origin, and Hills *et al.* (1974) recently applied several discriminant methods to a paleontological problem. Kim *et al.* (1966) discussed the problem of selecting characters for use in a discriminant function; much of the character-set optimization literature mentioned earlier is also relevant reading here.

Briefly, a discriminant analysis involves the weighting of characters by their ability to help distinguish between two taxa or two groups of taxa. If more than two taxa are being considered as possible identifications, a discriminant function for each pair is usually needed, or at least a dichotomizing series of discriminants eventually separating all the taxa (cf. Whitehead, 1954). Note that discriminant analysis does not tell whether an unknown belongs in a particular taxon or not. Rather, it helps tell in which of a pair of taxa the unknown is more likely to belong, assuming for the moment that it belongs to one of them. Other methods must be used to determine whether the unknown indeed belongs to either of the taxa being considered; techniques such as Hall's (1965) peculiarity index and Goodall's (1966) deviant index are helpful here, as is simple familiarity with the group of taxa being studied.

The methods of numerical taxonomy (Sneath and Sokal, 1973) also provide a powerful but expensive means of identifying difficult material, namely by including both unknown specimens and previously identified ones among the entities being classified by a clustering program. If an unknown is included in the same cluster as a named specimen, that taxon can be considered a tentative identification. However, if the unknown remains isolated, it is not similar to any of the other material included in the study, and deserves further attention. Campbell (1973) and Colwell *et al.* (1973) gave detailed presentations of such identification schemes, and similar procedures were presented by Brisbane and Rovira (1961) and by Bogdanescu and Racotta (1967). The implications of including nomenclatural types in such a study are discussed by Sneath and Cowan (1958) and in many more recent papers. Numerous instances of the implicit identification of unknown specimens by their inclusion in numerical-taxonomy studies are in the literature, but such analyses are rarely made solely for the purpose of identification. Indeed, numerical taxonomy becomes useful to identification only when the unidentified material provides significant additional knowledge about its taxonomic group, necessitating major realignments of taxon boundaries. Otherwise it should be sufficient to rank possibilities by similarities or to use discriminant functions, in either case using specimens identified according to the previous taxonomic treatment of the group as the reference for taxon descriptions and limits.

POLYCLAVES AND OTHER SEQUENTIAL METHODS

In sequential-polyclave identification procedures, characteristics of the unknown specimen are used one by one (or a few at a time) in a repeated process of elimination. First, an initial set of possibilities is considered. A number of characters are available for use, but particular characters are not specified *a priori* as in a dichotomous key. Instead, the user of a polyclave is free to select the characters he wants at each step in identifying an unknown specimen. Generally, these sequential procedures are called polyclaves or multiple-entry keys, although several other names have also been coined for them (cf. Morse, 1971, for further discussion). Since the user of a polyclave is free to select appropriate characters for each unknown specimen, the route taken to a particular name may differ considerably from one specimen to another. This freedom of choice in a polyclave is particularly valuable in attempts to identify incomplete or fragmentary material. Polyclaves also allow an experienced naturalist to achieve very rapid identifications by using characteristics that are shown by his specimen but generally rare or peculiar in the relevant group of taxa. However, even the most inexperienced beginner should be able to reach an identification eventually

B

with a polyclave, if identification of the specimen is possible at all with the available information.

Polyclaves can be either monothetic or polythetic in their operation. The critical difference between the two modes is whether a single difference between the specimen's description and the description of a particular taxon is considered sufficient evidence for eliminating that taxon from further consideration. With some polyclaves, the decision between monothetic and polythetic mode can be made as a user option for each specimen.

Three different kinds of polyclaves are currently in use. The simplest polyclaves are hand-held punched-card keys and various other overlay-card methods. Next in complexity come the mechanical methods, then computer-based polyclaves, including quantitative probabilistic methods. The role of computers in each of these basic implementations of the polyclave idea will be discussed next.

Field Keys and Mechanical Polyclaves

In a common kind of punched-card *field key*, each card represents one character state. Each taxon being considered as a possible identification is assigned a particular position on all the cards, and a hole is punched in that position on a card if that taxon shows the particular character state represented by that card. The key is used by selecting cards representing characters of the unknown specimen, aligning the cards, and noting any positions punched on all the selected cards. (Such positions can be found at once by holding the deck of cards up to a light source or a bright sky.) These positions indicate suggested identifications. Generally, such card decks are used in an exact-match, monothetic mode, since a difference on any single character is sufficient to block the light at some position. However, some light still comes through if there are only a few differences, so with practice such punched-card decks can also be used polythetically.

Most punched-card keys are hand-produced, at least in their originals; Hansen and Rahn's (1969) key to the families of flowering plants is a familiar example. Other kinds of hand-held polyclaves, such as the overlay sheets developed by Duke (1969), are also hand made. Recently, Weber and Nelson (1972) used standard data-processing cards and equipment to produce a punched-card key to the genera of mosses in Colorado. They punched the original of each card at a keypunch machine, assembled a master deck, then replicated this deck many times on a reproducing card punch to produce multiple copies of the key with minimal effort. Leila Shultz simplified this procedure further by punching the master deck by computer program from a taxonomic data matrix; her key to

Colorado wildflowers (Shultz, 1973) was produced by this automated method. Peters (1972) also saw this possibility for automation, and the program I wrote with him for card-key production is included in my recently-published program package (Morse, 1974a). Pankhurst and Aitchison (this volume, p. 73) also developed such a program, and yet others may exist.

Keys using edge-punched cards have also been developed for specimen identification (e.g. Clarke, 1938, Normand, 1946). Here each card represents a taxon, and the characters and character states are assigned particular punched positions around the edges of the cards. When a taxon can be expected to show a particular character state, the edge of the card is notched for that character; otherwise the hole is left intact. The deck is used by aligning the cards, selecting a character of the specimen, inserting a stiff wire or needle through the hole for that characteristic on all the cards, and shaking the deck slightly so cards notched at that position drop out while the others are held by the needle. The process can then be repeated using other characters; at each step, the cards which drop are those of taxa showing the selected character. An alternate, more efficient method for using edge-punched cards is to notch the cards of taxa *lacking* the specified character, so these will drop out of the deck when that character is selected. The remaining cards represent taxa showing the selected character, and can be used immediately for further selection with a different character. Neither implementation of the edge-punched card key can be used polythetically without considerable difficulty, so edge-punched cards are appropriate only for problems in which the taxa are reasonably distinct. They do offer one distinct advantage over the face-punched cards discussed earlier, namely that the cards need not be kept in any particular order for convenient use.

Edge-punched cards cannot be notched directly by any computer equipment I am aware of, but a computer might be used indirectly in preparing such a key. Little (1968) has developed a concise means of presenting instructions for making such a key by hand; numerical descriptions of this sort could readily be produced by computer from data matrices.

Mechanical polyclaves range in complexity from Ogden's (1953) and Leenhouts' (1966) elegantly simple pencil-and-paper methods to the "information sorter" developed by Olds (1970), with its red and green lights. The topic of computer involvement in the development of mechanical polyclaves has received little attention, aside from the general work in character-set optimization mentioned earlier. Programs could be readily written for such purposes as printing Leenhouts-style keys from taxonomic data matrices. Also, ranked listings of characteristics by within-group peculiarities could be produced to help inexperienced polyclave users develop an observing strategy.

Computer-based Polyclaves

Two general kinds of computer-based polyclaves are in common use. One kind, developed for qualitative taxon/character data, employs threshold elimination, while the other, developed for taxon/character frequency tables, employs likelihood ratios or other probabilistic techniques. In either case, an initial list of possible taxa is gradually reduced as information about the unknown specimen is supplied.

Frequently, a computer is used on-line in an interactive mode to support a polyclave program. This permits the user to submit observations one by one, or a few at a time, and immediately obtain such results as a revised list of possible taxa. He can then use this new information as a basis for selecting the next characters to study. With some programs, he can even ask the computer to suggest appropriate characters to observe next, considering the current situation. Although they are expensive at present computer-service prices, on-line interactive polyclaves offer several advantages over one-pass, batch-processing polyclaves. These advantages include rapid results, efficient character selection, minimal observation, and early error correction. Figure 2 shows the kind of dialog that can be developed with an interactive polyclave.

Most present on-line polyclaves use character codes or numbers to communicate the specimen's description to the computer. A text processor can also be employed here, as shown by Mullin (1970), but this approach not only takes more computer time, but also slows down the dialog since significantly more typing is involved. The use of key words from a controlled vocabulary may be a better approach; simple rules of syntax similar to those for traditional diagnostic descriptions could be developed. Polyclaves can also be developed using interactive graphics, as shown by Boughey *et al.* (1968); Pankhurst's (1972) data-capture procedures are similar. Martin (1973) has provided a thorough discussion of communication problems in interactive systems.

The on-line polyclave is particularly powerful as an aid to teaching specimen identification, a possibility explored extensively at Michigan State University in recent years (Beaman, this volume). Grimes *et al.* (1972) also describe an interesting teaching application using the polyclave idea. Here they simulate bacteriological identification, letting the student select laboratory procedures and tests for identifying a hypothetical unknown bacterium. A dollars-and-cents cost is assigned to each test and procedure available, and the student's goal is to minimize the total cost of identifying the unknown pathogen. Their implementation of this program on the PLATO system at the University of Illinois is particularly interesting for its use of computer graphics.

While on-line polyclaves are a powerful identification tool, batch-processing

```
*****   M S U   TAXONOMIC DATA MATRIX PROGRAMS   *****

WHICH MSU PROGRAM?IDENT4

DATA MATRIX?ACER1M

FILE <ACER1M>, REVISED 03/01/70

        MICHIGAN MAPLES MATRIX  --  --  BY LARRY MORSE, MARCH 13, 1969
           BASED ON VARIOUS FLORAS AND MANUALS AND SPECIMENS IN MSC
ANY EDITING?NO


IDENTIFICATION PROGRAM:   1=INSTRUCTIONS   2=START --?1


    PROGRAM <IDENT> FOR DIRECT IDENTIFICATION OF SPECIMENS USING
TIME-SHARING COMPUTERS AND TAXONOMIC DATA MATRICES.  PREPARED BY
LARRY MORSE, MICHIGAN STATE UNIVERSITY.  THIS EDITION LAST REVISED
30 MAY 1973.

OPTIONS FOR REPLY TO NEXT  --
    0=SUGGEST USEFUL CHARACTERS
    1=CONTINUE WITH SAME SPECIMEN
    2=LIST POSSIBILITIES REMAINING
    3=DELETE LAST CHARACTER-INPUT SET
    4=START NEW SPECIMEN, SAME MATRIX
    5=REQUEST NEW MATRIX FOR NEW SPECIMEN
    6=RECYCLE PROGRAM FOR NEW USER
    7=STOP PROGRAM RUN

***** NEW SPECIMEN:   COLLECTOR AND NUMBER --?M. B. MOORE 396
VARIABILITY LIMIT?0
CHARACTERS?21,53,71
SUGGESTED IDENTIFICATION OF SPECIMEN M. B. MOORE 396
        A. SACCHARINUM
    UNUSUAL CHARACTERS:        71       111       143       173       13
    CHARACTERS YOU USED:       21        53        71
NEXT?4


***** NEW SPECIMEN:   COLLECTOR AND NUMBER --?L. E. MORSE 6431
VARIABILITY LIMIT?1
CHARACTERS?31,171,61
THE FOLLOWING  2 OF  10 TAXA REMAIN:
        A. PLATANOIDES
    -1  A. NEGUNDO
NEXT?1
CHARACTERS?13
SUGGESTED IDENTIFICATION OF SPECIMEN L. E. MORSE 6431
        A. PLATANOIDES
    UNUSUAL CHARACTERS:        31       171        43       111       131
    CHARACTERS YOU USED:       13        31        61       171
NEXT?4
```

FIG. 2. Dialog generated during identification of two specimens using computer-based
polyclave procedure developed at Michigan State University. The program
(Morse, 1974a) is here used with a small sample data matrix in identifying two
specimens of maple (*Acer*) from Michigan. Most of the text shown here was
printed by the computer; user input follows question marks. The first specimen
was identified quickly using the program's monothetic mode, while the second
was identified in two steps using the polythetic mode with a variability limit of
one. A numbered list of characters was kept at the computer terminal in the
herbarium, so only code numbers are used in the "conversation" with the
computer.

polyclaves have also been developed. Here, of course, only one set of characteristics per specimen can be submitted per pass, resulting in a considerable delay (often hours or days) if the submitted characters are insufficient to identify the specimen. At first glance, a single-pass polyclave looks like one of the character-set identification methods discussed earlier, but there is the important distinction that for a character-set method, information on all the characters must be supplied, while for a polyclave, any subset of those characters can be tried. With a polyclave, a few well chosen characters can perform much better than a large number of poorly chosen ones. However, in general one cannot know in advance whether a particular small set of characters will indeed lead to an identification. Accordingly, more characters are generally used in a batch-mode polyclave than would be needed to make the same identification interactively. On the other hand, the batch-processing implementation of a polyclave is presently significantly less expensive to use than an interactive version, in terms of computer costs. Thus, batch-processing polyclaves may be most appropriate for applications such as identification in microbiology, where the cost of using a few too many characters in the first try is minor compared to the cost of waiting for additional tests to be done for a second try at the identification.

A simple batch-processing implementation of the polyclave idea involves the use of "generalized information systems" to pose specific queries to a large data base (Krauss, 1973). Keller and Crovello (1973) have used a floristics data base in this way, and Boughey *et al.* (1968) provided a detailed description of their procedures and programs for specimen identification with such a data system. The TAXIR system (Rogers, 1970; Estabrook and Brill, 1969) can be readily used for specimen identification, as can the detailed palynological data bases of Walker *et al.* (1968) and of Germeraad and Muller (1970, 1974). However, most of these generalized procedures are cumbersome for routine work, as Shetler (1974) has concluded about GIS from his experiences with "Flora North America", so a number of special-purpose programs for specimen identification have been written to improve the accessability of data-bank information in identification work. An important goal in much of this programming work has been to release the user of the program from any need to know computer programming himself, thus making computer-assisted identification readily accessible to the minimally trained student or technician, as well as to the senior researcher who has no personal desire to learn the intricacies of computer programming.

Several qualitative-data programs for specimen identification have been developed in recent years. These include the fairly generalized programs of Boughey *et al.* (1968) and Mullin (1970), as well as the more finely tuned programs by Goodall (1968), Morse (1969, 1971, 1974a; Beaman, this volume, p. 267) and Pankhurst and Aitchison (this volume, p. 181). Goodall's is highly conversational, but seems to involve too much text for use with present typewriter-like terminals such as a teletype; his approach may be practical with a television-like terminal of some kind, such as the one used by Kendrick (1972) for a computer-stored key. Pankhurst and Aitchison's program also prints long lists of character choices at the terminal. My program uses compact character codes instead, but requires that various character lists be kept at the terminal for reference during use of the program.

Conceptually, all these monothetic polyclave programs operate by a simple elimination procedure. After an initial list of possible taxa is selected, and tabular information on their characteristics is read into the computer memory, the identification program enters a three-step cycle which is repeated each time the user wants to supply more information about the observed characters of his unknown specimen. In this identification cycle, the computer first accepts data on one or more characteristics of the specimen, then eliminates from the list of possibilities any taxa whose data disagree with these specimen data, and finally takes appropriate action based on the results of the elimination step. Cycle-ending actions include printing the name of a suggested identification, printing a message that no possibilities remain, printing a short list of remaining possibilities, and printing a message stating how many possibilities remain. The user can then base his next action on these results, perhaps continuing, back-tracking a step to correct an error, or starting over with the same or another specimen. It is this close interaction of researcher and computer that makes the interactive sequential methods such powerful and efficient identification procedures, both for research and museum use and for teaching.

Polythetic polyclave procedures have been developed from monothetic methods by Morse (1971), and from conditional-probability methods by Dybowski and Franklin (1968) and by Lapage's group (Lapage *et al.*, 1970, 1973; Bascomb *et al.*, 1973; Willcox *et al.*, 1973; Willcox and Lapage, this volume), as well as by numerous workers in other fields.

The nonprobabilistic polythetic polyclave is readily implemented by making one modification to the general monothetic-polyclave algorithm outlined above. Specifically, taxa are not simply eliminated from the list of possibilities, but instead a tally is kept for each taxon of the number of times a disagreement between the specimen and that taxon is found. At the end of each identification cycle, the taxa are ranked by this tally, and those having few or no differences with the specimen are printed. In my program (Morse, 1974a), I simply allow the user to set a threshold or "variability limit" specifying the number of differences which the program will tolerate before eliminating a taxon from the set of remaining possibilities. At present I know no objective criterion for setting this variability limit algorithmically; presumably the limit depends on the variability of the taxonomic group and the reliability of the data, as well as the condition of the specimen, the importance of accuracy in the identification, and the user's experience in observing taxonomic characters. We typically use a variability limit of one or two for well studied groups of vascular plants, but have little experience in other groups. A floating limit might also be used, along the lines of Gyllenberg and Niemelä's concept of "neighborhood" discussed earlier. Our general experience with polythetic polyclaves shows they are powerful tools for identification in difficult situations, but require that substantially more characters be examined for each specimen than in making monothetic identifications.

Probabilistic polyclaves in biology generally employ the "conditional probability" or "likelihood" approach to decision-making, although other statistical models (such as Bayes' theorem methods) might also be used (cf. Ledley and Lusted, 1959a).

To use the likelihood model of identification in some situation, one must first know the likelihood (commonly expressed as a frequency or percentage) of obtaining a given test result (or observing a given characteristic) for each of the taxa being considered as possible identifications. Furthermore, the model assumes the tests are independent, so only one test out of each set of dependent or correlated characteristics can be used. Assuming the characters are all independent, the likelihood of obtaining specified results for two tests for a given taxon can be found simply by multiplying the tabulated likelihoods of obtaining each result separately. For example, if 95% of the *Escherichia coli* isolates sampled in a particular study gave a positive indole test, and 95% gave a negative gelatin test, then it is assumed in the likelihood methods that 0.95×0.95 ($= 90.25\%$) will give both a positive indole test and a negative gelatin test, and so forth; such predictions are true only if the characters involved are not correlated.

In *maximum likelihood* identification, several characters of the unknown are first determined. Then, for each taxon under consideration, the likelihood is calculated that that taxon shows the characters in the same states as observed for the unknown specimen. The taxon having this likelihood calculation highest is then suggested as a possible identification. In practice, the *ratio* of likelihoods of the most likely and the next most likely possibilities is also important to decision-making, for this ratio is high only if one possibility is distinctly more likely than any other. To allow for variation within a taxon, Dybowski and Franklin make the important adjustment of dividing each taxon's likelihood by the maximum possible for that taxon. They then rank the various taxa by decreasing values of this "modal likelihood fraction". For deciding between outright identifications and cases where several taxa remain likely, the Lapage group calculates an "identification score" for each of the possible taxa. They do this by normalizing the various likelihood scores for the various taxa so that these scores total unity. They then accept a particular taxon as an identification only if its normalized likelihood score exceeds 0·999, implying the scores for all other taxa are below 0·001. If a satisfactory identification is not reached with the given information, sets of characters for additional investigation are listed by the Lapage group's program.

Identification methods using Bayes' theorem are similar to those using likelihood considerations, except that the various taxa are weighted *a priori* by the likelihoods that the users of the identification system will encounter each of them in their work; for details see Good (1959) or an advanced statistics book. Basically, the effect of this taxon weighting is to reduce the identification scores of rare taxa, wherever they occur in the list of possibilities, therefore removing rare taxa from the top area of the list if there are also any commoner taxa there. At the least, I believe the advisability of doing such a thing in critical work is questionable; if anything, we use computer-assisted identification to remind us what rare taxa we may be overlooking as possibilities! Nevertheless, the Bayesian approach has been used to some extent in medical diagnosis (e.g. Boyle *et al.*, 1966; Gorry and Barnett, 1968; Lusted, 1968) and other areas, but the only specimen-identification system employing it that I know of is Baum and Lefkovitch's (1972) for cultivars of oats.

In any case, the lack of data on local taxon abundances (actually expected encounter frequencies) necessary for Bayesian identification makes this method impractical for most applications in natural history (Morse, 1971). For example, Hawkes *et al.* (1968) discussed the work needed to obtain and process such information for vascular plants of a single British county, and King *et al.* (1967) showed the even greater complexity of quantitative data on seabird sightings. Webb *et al.* (1967), in an ecological paper, noted that the sheer bulk of the site/species data potentially available in tropical forests strains both their observational powers and their computational resources, as should be obvious to any field naturalist thinking of measuring local taxon abundances over any respectable area. However, the identification of medical bacteria is an important exception to this general pessimism, since large quantities of comparable data are easily gathered, although these data must always be approximations (cf. Ledley, 1969; Gustafson *et al.*, 1969). Abundance data of similar quantity might also be obtainable for other widely observed kinds of organisms, such as plants and animals of economic importance, pests of crop plants and of domestic animals, and conspicuous wild species widely observed by the public, especially birds, butterflies, and flowers.

Routine, fully automated specimen identification may be possible for at least some kinds of organisms through the combination of automated character observation and computer-based pattern recognition procedures. Automated character observation would provide a concise, digitalized description of certain characteristics of the specimen. This digital profile can then be compared by some means with various reference profiles by a pattern recognition program, which can then suggest the identification if one is reached. In the ideal implementation, a technician would insert a sample of the unknown specimen in the machine at one end, then moments later receive the identification and supporting information from a printer at the other end of the apparatus. Some promising methods of specimen observation will be discussed briefly here, but adequate consideration of appropriate programs for data analysis is impossible in this survey. Various programs being developed for analysis of chemical and astronomical spectra may be particularly appropriate for study of the profiles produced by some of these automated devices; Clerc and Erni (1974) review some of these methods. There is, of course, already a vast literature on automated recognition of digitalized visual patterns, recently reviewed by Rosenfeld (1973). Barlow et al. (1972) have provided a fine introductory discussion of the challenges of this field.

Two techniques that have been explored in some depth in recent years for automated character observation are optical scanning and pyrolysis–gas–liquid chromatography. Several other techniques of possible use in automated identification have been investigated with varying success, including infrared spectrophotometry (Riddle et al., 1956), chromatography followed by mass spectroscopy (Meuzelaar and Kistemaker, 1973), neutron-activation analysis (Hess et al., 1968), fluorescent antibodies (Preece, 1968), and microcalorimetry (Boling et al., 1973). However, not all these can be expected to prove useful or economical in diagnostic work (cf. Steel, 1962). For conventional chromatograms and stained electrophoretic gels, pattern-recognition procedures might be used in identification work providing sufficient standardization and reproducibility could be achieved (cf. Runemark, 1968), although this uncertainty may be no greater than for conventional morphological characters (Adams, 1974).

In a study of the potential value of optical scanning for characterizing biological specimens, Rohlf and Sokal (1967) reduced line drawings of mosquito pupae to coarse binary images, and used these data as characters for a clustering program. They also discussed such problems as standardization and alignment involved in applying direct scanning to real specimens rather than just drawings of them, itself a difficult enough problem (Freeman, 1974). Chromosome analysis by the scanning of photomicrographs is now well advanced (e.g. Ledley and Ruddle, 1966; Rutovitz, 1968; Gilbert and Muldal, 1971), as is recognition of abnormal cells in cytological samples (e.g. Rosenberg *et al.*, 1969; Bartels *et al.*, 1970); the success of these studies suggests that similar methods might be employed for identifying organisms or organism parts having distinctive two-dimensional shapes. However, extension of these direct-scanning methods to the identification of diverse large organisms may be impractical, because of the difficulties introduced by three-dimensional form. For most taxonomic groups, it may be more useful to explore computer techniques for obtaining some of the information needed in traditional identification methods (e.g. Macinnes *et al.*, 1974).

Flat objects such as leaves or insect wings may serve as better subjects for present pattern-recognition methods (e.g. Belson and Dunn, 1970). Colonies of bacteria can also be treated as two-dimensional objects for pattern-recognition purposes. Glaser and Ward (1972) demonstrate the feasibility of automated identification of bacteria by colony morphology using advanced pattern-recognition procedures. They suggest their methods might be in large experimental studies or screening projects, and also suggest their technique could be combined with replicate plating using different culture conditions to provide better differentiation of taxa having similar colonies. Automated devices using pattern recognition to monitor the growth of microbe cultures in various culture media have also been developed (e.g. Gerke *et al.*, 1960; Glaser and Wattenburg, 1966; Bowman *et al.*, 1967); used with carefully selected sets of growth media and linked to a computer for direct analysis, such devices could provide fully automated bacterial identification.

For most groups of organisms, analysis of a specimen or sample by pyrolysis or some other technique appears more promising as a basis for automated identification than does direct imaging. In the pyrolysis methods, a sample of the specimen is heated intensely in the absence of air, and the resulting degradation products are chemically characterized, usually by gas–liquid chromatography or by mass spectroscopy. Success with this method is reported for a diversity of organisms, including bacteria (Reiner, 1965; Reiner and Ewing, 1968; Meuzelaar, 1974), algae (Nichols *et al.*, 1968; Sprung and Wujek, 1974), fungi (Vincent and Kulik, 1970, 1973; Seviour *et al.*, 1974), insects (Hall and Bennett, 1973), and even plant pathogens still in the host tissues (Myers and

Watson, 1969). Since the material is destroyed by pyrolysis, it is inappropriate
for small museum specimens, but could become important for use with small
tissue samples of large specimens as well as for such problems as identification in
large ecological surveys, where ample fresh material is generally available. When
coupled to appropriate computer programs for data analysis and comparison
(e.g. Landowne *et al.*, 1972; Menger *et al.*, 1972), pyrolysis offers considerable
promise for fully automated identification of almost any biological materials,
providing the necessary reference pyrograms can be developed. Pyrolysis may
be especially important for groups such as microorganisms where visible
characteristics are inadequate for precise identification. The newer techniques
combining pyrolysis and mass spectroscopy, rather than chromatography, may
make extremely rapid identification of bacterial colonies possible, since the
analysis takes less than a minute to complete. Heller's (1972) polyclave program
for mass-spectrum data deserves mention here.

Another approach to automated identification involves pattern recognition
from remote-sensing information. Such methods as visible-light and infrared
photography from aircraft or satellites provide the data here, which is then
analyzed in any of numerous ways. These procedures are by now well developed
for use in geography, geology, and vegetation analysis (Colwell, 1973), but so
far as I am aware they have been applied to specimen identification only in the
sense of recognizing wild species or crop plants in pure stands during ecological
or agricultural surveys. Vlcek (1972) provided a discussion of the data-collecting
and data-storage aspects of the use of remote sensing in such vegetation studies.

DISCUSSION

The broad range of algorithms now developed for computer-assisted specimen
identification is clearly shown by the numerous examples discussed in this survey
and throughout this volume. Which method might be used in a given situation,
or indeed whether computer use is appropriate at all, depends on the particular
requirements of the individual project. However, it is presently difficult to make
any general comparisons of all these identification methods, since they have been
developed for different purposes and tested with different data.

Need for Comparative Studies

Several interesting experiments could be designed to compare the suitability of
various identification aids for particular needs, such as identification of medical
bacteria, specimen-sorting in museums, rapid identification in ecological studies,
or field use by amateurs. Thorough discussions of the use of these various new
identification methods by inexperienced users are also welcome contributions to

our understanding of their capabilities and limitations. Watson and Milne (1972) presented such a critique of Pankhurst's key-constructing program, and Gilmartin (1970), Stevens (1971), Taggart (1971), Furlow and Beaman (1971) and Hanks (1972) have discussed their experiences with early versions of my programs. Detailed discussions comparing experiences with both modern and traditional methods for particular problems are also needed. Whiffin (1973) used four methods for studying a hybrid swarm, and found each made unique contributions to his understanding of the problem. Gipson *et al.* (1974) also show the usefulness of alternative quantitative techniques in a geographical study of hybridizing taxa. Bascomb *et al.* (1973) compared traditional and probabilistic-polyclave methods for microbiological identification. Hansen and Cushing's (1973) study of methods for fossil pollen identification is a particularly useful discussion of difficulties encountered (and mostly overcome) in using computer-based procedures to improve the quality of identifications in their field.

Future Role of Computers in Identification

Another area where more information is needed concerns the rates of obser-vational errors for various kinds of taxonomic characters, various groups of organisms, and various degrees of experience of the observers. Two such studies have been done by Sneath and Johnson (1972) and Adams (1972), but far more attention to this topic is needed before we can make rational choices between alternative identification procedures for particular problems.

Despite the lack of detailed comparisons in many areas, some generalizations about the potential value of computer-based aids to specimen identification can be readily made. For traditional monographic and revisionary taxonomic studies, as well as the preparation of field manuals and similar works, computer aids to editing and constructing dichotomous keys and to preparing diagnostic descriptions should be helpful. Computer-produced punched-card keys may become common for office use, but are bulky in their present forms for general use in the field. The powerful probabilistic methods are helpful in clinical microbiology where the required data are available, but seem too demanding for use in other areas where simpler methods will suffice. Fully automated pattern-recognition systems are important possibilities in certain specialized applications, such as microbial identification, large-scale ecological or agricultural surveys, and identification of difficult fragmentary material. Computer-supported polyclaves may be most important for sorting centers, large museums, and oceanographic surveys where mixed materials from diverse groups are initially identified. Such tools could also help museum curators or their assistants make routine service identifications, especially for large lots of material. The powerful character-set methods such as discriminant analysis may prove worthwhile for use only in very difficult situations where conventional identification is not possible, such as problems involving sibling species, hybrid swarms, apomicts, cultivars, or palynological samples.

On the other hand, it is important to remember that the new identification aids discussed at this symposium need not be used every time a specimen is identified. As Mayr (1971) correctly concluded, most specialists already familiar with their groups have little use for computer methods—whether for classification or identification—in day-to-day work with their material, except when such methods as discriminant analysis are called for to help differentiate closely related or apparently overlapping taxa. Printed dichotomous keys have served remarkably well over the years as identification aids for many groups, and where keys are still appropriate, computers are better used, if at all, in helping write better keys than in replacing the printed keys with more demanding and more expensive methods. However, the younger generation of systematists, more comfortable with computers than their academic elders, may try computer techniques readily whenever they encounter difficult identification problems.

For amateurs and the public generally, the most important benefit of computer-assisted specimen identification will be indirect, namely in improving the quality of the identifications upon which the entire information system of taxonomy depends (Morse, 1970). Computer-constructed keys and computer-produced descriptions and comparisons may appear in popular field guides, but the casual reader would not be aware that a computer had been used in their production. I doubt that punched-card keys will gain much public acceptance, however useful they might prove to professional systematists and naturalists. Direct use of simple, interactive identification programs by the public is feasible at costs comparable to those of other major museum exhibits (Peters, 1972). Such implementations of computer-assisted identification could both help increase the public's awareness of the biotic world about them, and help relieve museum personnel from some routine identification work. The educational use of computer-assisted specimen identification in biology courses serves similar purposes, educating the student both in the process of identification and in the characters and names of the specimens he is working with (Beaman, 1971 and this volume). However, instruction in the use of traditional printed keys should not be neglected in such courses; Beaman's (1971) "Syllabus" shows the balance that can be reached. Crovello (1974) presents an excellent introductory survey of possible roles of computers in biological education.

Perhaps the most important use of these new identification techniques will be in identifying fragmentary or incomplete materials and sterile or immature forms, now frequently regarded as essentially unidentifiable. Such increased identification capabilities may be particularly useful in paleontology, ecology, agriculture, and public health (cf. Sokal and Sneath, 1966; Meijer, 1967; Williams, 1967; Duke, 1969; Michener *et al.*, 1970; Peters, 1972; Pettigrew and Watson, 1973). These procedures may be particularly useful in palynology, and for identification problems involving sibling species.

Well Organized Data Required

Fundamental to any discussion of computer-assisted specimen identification is an understanding of the data requirements of various methods. For efficient operation, most of the techniques described here require a full "taxonomic data matrix" giving taxon/character entries for each of a respectable number of characters for all the taxa being considered in the study. Elsewhere, I recently presented a general discussion of the computerization of taxonomic information (Morse, 1974b), but some aspects of this topic deserve repetition here. A major requirement of identification systems is that the information on a given characteristic be recorded and encoded in the same way for all the taxa being studied. This problem of *data comparability* is a serious limitation to the development of information systems in biology, particularly synoptic ones based primarily on previously published taxonomic descriptions. Ainsworth (1941) faced this problem in developing a polyclave to fungi three decades ago, and Sneath and Cowan (1958) discussed their difficulties in obtaining comparable data for a high-level numerical taxonomy study of bacteria. More recently this problem has been discussed by Crovello (1968), Colwell and Wiebe (1970), Gilmartin (1970), Furlow and Beaman (1971), Keller and Crovello (1973), Shetler (1974) and by Pankhurst elsewhere in this volume (Chapter 14).

Ideally, for identification work one should have strictly comparable data for a large number of characteristics available in machine-readable form for all the relevant taxa. However, most computer programs for specimen identification investigated to date perform surprisingly well on strikingly incomplete data bases (Furlow and Beaman, 1971; Pankhurst, this volume, p.87), with the obvious exception of the character-set programs. Polyclave programs are particularly tolerant of omissions in the data base, although a large number of missing entries slows the identification process considerably. Yet tolerable performance of such programs on poor data is no excuse for not gathering better data when obtainable, and the probabilistic data matrices developed recently by Lapage's group for medical bacteria show what can be done with a simple character set and a large number of samples. Many of the detailed data matrices developed for numerical taxonomy should also be directly adaptable to identification.

However, when morphological characters of higher taxa are considered, the problems of comparability, convergence, homology, and verbal description of the features become challenges of their own (cf. Turrill, 1957; Kendrick, 1965; Dale, 1968; Williams, 1969; McNeill, 1972; Rogers and Fleming, 1973; Porter *et al.*, 1973; Morse, 1974a, b). As noted earlier, problems of characterization are especially difficult when most or all of the information is derived from the literature alone; ideally, specimen identification should employ information ultimately traceable to observations of previously identified museum specimens!

General discussions of the need for well-organized basic taxonomic data are now frequently made.* The need for better taxonomic information is now widely recognized, and taxonomic data banks of various kinds are being promoted as a solution to this problem. However, the systematist must recall that even the most carefully organized basic data matrix inevitably involves some information loss when compared with original observational data on particular specimens. For most purposes, this loss is trivial compared with the gain in data accessability offered by an information system, but the computer is no replacement for the descriptive literature and the museum collection. Careful design is most important in developing a data system for taxonomic information, for the data potentially available far exceed the capacity of any information system now feasible or desirable. Heywood (1973), Morse (1974b), and Shetler (1974) have discussed some of these design choices recently.

Theoretical Advances not Computer-dependent

Finally, it is important to remember that two rather different subjects are intimately intertwined in the recent work on computer-assisted specimen identification. In recent years we have made both remarkable advances in the theory of specimen identification, and fascinating computer-based implementations of many of these new ideas. The theoretical advances do not themselves depend on computer equipment; with sufficient time and scratch-paper, any of the new methods can be used by hand. The same is true of numerical taxonomy and other "computer" procedures—without computer support, they are only impractical, not impossible (cf. Knuth, 1973). Computer-based identification systems are only as good as the theory, programs, reference data, and specimen observations they employ; the mere use of a computer in a project is no assurance whatsoever of objectivity or accuracy, as Mayr (1965), Rollins (1965), and Sneath (1971) have clearly stated. The new identification procedures

* Stands representative of this view have been taken recently by Cutbill (1971b); Morse, Peters and Hamel (1971); Shetler (1971); Watson (1971); Peters (1972); Gómez-Pompa and Nevling (1973); and Morse (1974b).

perform as they do, not because they employ fancy, magical equipment, but because they consider a broad range of characters and make identifications polythetically, on the basis of overall similarities rather than possession of a handful of "key characters". Perhaps the most important contribution of the research reviewed here has been the heightened awareness of the needs to use large numbers of characters and to avoid arbitrary monothetic methods in biological specimen identification, regardless of whether computers are or are not used in a particular application.

CONCLUSIONS

In recent years, a number of novel procedures for specimen identification have been proposed and developed; most of these require computer support for practical use. Algorithmic key construction and direct identification of unknown specimens by character-set and polyclave procedures have received the most attention in this work. During this period of development, the polythetic taxon concept has gained wide acceptance among systematists, along with the corollary implication that the traditional dichotomous key is inappropriate for many specimen-identification situations. Also, diverse new kinds of taxonomic characters have become available in recent years for use in identification systems as well as in taxonomic studies.

A few fully automated identification systems are being developed for specialized applications, but most computer-based identification procedures rely on a human observer to examine the unknown specimen and state some of its characteristics. Considerable progress has been made recently in the development of polyclave procedures, which allow the user to select the characteristics to observe for each specimen being identified. Several simultaneous character-set methods have also been developed, primarily for identification in microbiology. Powerful statistical methods such as discriminant analysis are available for challenging problems, but are rarely used because of their cost and complexity.

In the past decade there has thus been a vast expansion of our awareness of the alternative approaches to specimen identification theoretically available. However, actual, everyday use of computers in routine specimen identification is still quite rare. Also, dichotomous keys and similar methods are adequate for many identification problems, and need not be replaced for groups where they are satisfactory. Instead, novel computer methods for selecting characters and constructing keys may improve the quality of these traditional tools. Computer-based procedures should be most helpful in two areas, the differentiation of very similar taxa and the initial assignment of unknowns to major groups, such as families, orders, classes, and phyla.

42   *L. E. Morse*

REFERENCES

ADAMS, R. P. (1972). Numerical analysis of some common errors in chemosystematics. *Brittonia* **24**, 9–21.

ADAMS, R. P. (1974). Numerical chemotaxonomy. II. On "Numerical Chemotaxonomy" revisited. *Taxon* **23**, 336–338.

AINSWORTH, G. C. (1941). A method for characterizing smut fungi exemplified by some British species. *Trans. Br. mycol. Soc.* **25**, 141–147.

ANDERSON, E. (1949). "Introgressive Hybridization". Wiley, New York.

BAER, H. and WASHINGTON, L. (1972). Numerical diagnostic key for the identification of Enterobacteriaceae. *Appl. Microbiol.* **23**, 108–112.

BARLOW, H. B., NARASIMHAN, R. and ROSENFELD, A. (1972). Visual pattern analysis in machines and animals. *Science* **177**, 567–575.

BARNETT, J. A. (1971). Selection of tests for identifying yeasts. *Nature New Biol.* **232**, 221–223.

BARNETT, J. A. and PANKHURST, R. J. (1974). "A New Key to the Yeasts". North-Holland, Amsterdam and London; American Elsevier, New York.

BARTELS, P. H., BAHR, G. F., BELLAMY, J. C., BIBBO, M., RICHARDS, D. L. and WIED, G. L. (1970). A self-learning computer program for cell recognition. *Acta Cytol.* **14**, 486–494.

BASCOMB, S., LAPAGE, S. P., CURTIS, M. A. and WILLCOX, W. R. (1973). Identification of bacteria by computer: Identification of reference strains. *J. gen. Microbiol.* **77**, 291–315.

BAUM, B. R. and LEFKOVITCH, L. P. (1972). A model for cultivar classification and identification with reference to oats (*Avena*). II. A probabilistic definition of cultivar groupings and their Bayesian identification. *Can. J. Bot.* **50**, 131–138.

BEAMAN, J. H. (1971). "Introductory Plant Systematics: Lecture Syllabus and Laboratory Manual". (4th edition). Department of Botany and Plant Pathology, Michigan State University, East Lansing.

BEDZEK, J. C. (1974). Numerical taxonomy with fuzzy sets. *J. Math. Biol.* **1**, 57–71.

BELLMAN, R. E. and ZADEH, L. A. (1970). Decision-making in a fuzzy environment. *Management Sci.* **17**, B141–B164.

BELSON, M. and DUNN, R. A. (1970). Pattern recognition of leaf contours. (Abstract). *In* "IEEE Conference Record of the Symposium on Feature Extraction and Selection in Pattern Recognition", p. 68.

BISBY, F. A. (1970). The evaluation and selection of characters in angiosperm taxonomy: an example from *Crotalaria*. *New Phytol.* **69**, 1149–1160.

BOGDANESCU, V. and RACOTTA, R. (1967). Identification of mycobacteria by overall similarity analysis. *J. gen. Microbiol.* **48**, 111–126.

BOLING, E. A., BLANCHARD, G. C. and RUSSELL, W. J. (1973). Bacterial identification by microcalorimetry. *Nature, Lond.* **241**, 472–473.

BOSSERT, W. (1969). Computer techniques in systematics. *In* "Systematic Biology", pp. 595–614, publ. 1692, National Academy of Sciences, Washington, D. C.

BOUGHEY, A. S., BRIDGES, K. W. and IKEDA, A. G. (1968). An automated biological identification key. *Mus. Syst. Biol., Univ. Calif., Irvine, Res. Ser.* no. 2, 1–36, i–xix.

BOWMAN, R. L., BLUME, P. and VUREK, G. G. (1967). Capillary-tube scanner for mechanized microbiology. *Science* **158**, 78–83.

BOYLE, J. A., GREIG, W. R., FRANKLIN, D. A., HARDEN, R. McG., BUCHANAN, W. W, and McGIRR, E. M. (1966). Construction of a model for computer-assisted diagnosis: Application to the problem of non-toxic goitre. *Q. Jl Med.* **35**, 565–588.

BRISBANE, P. G. and ROVIRA, A. D. (1961). A comparison of methods for classifying rhizosphere bacteria. *J. gen. Microbiol.* **26**, 379–392.

CAMP, W. H. (1945). The North American blueberries with notes on other groups of Vacciniaceae. *Brittonia* **5**, 203–275.

CAMPBELL, I. (1973). Computer identification of yeasts of the genus *Saccharomyces*. *J. gen. Microbiol.* **77**, 127–135.

CLARKE, S. H. (1938). The use of perforated cards in multiple-entry identification keys and in the study of the inter-relation of variable properties. *Chron. Bot.* **4**, 6, 517–518.

CLERC, T. and ERNI, F. (1973). Identification of organic compounds by computer-aided interpretation of spectra. *Fortschr. chem. Forsch.* **39**, 91–107.

COLWELL, R. N. (1973). Remote sensing as an aid to the management of earth resources. *Am. Scient.* **61**, 175–183.

COLWELL, R. and WIEBE, J. (1970). "Core" characteristics for use in classifying aerobic, heterotrophic bacteria by numerical taxonomy. *Bull. Georgia Acad. Sci.* **28**, 165–185.

COLWELL, R. R., LOVELACE, T. E., WAN, L., KANEKO, T., STALEY, T., CHEN, P. K. and TUBIASH, H. (1973). *Vibrio parahaemolyticus*—Isolation, identification, classification, and ecology. *J. Milk Fd Technol.* **36**, 202–213.

CORLETT, D. A. Jr., LEE, J. S. and SINNHUBER, R. O. (1965). Application of replica plating and computer analysis for rapid identification of bacteria in some foods. I. Identification scheme. *Appl. Microbiol.* **13**, 808–817.

COWAN, S. T. and STEEL, K. J. (1960). A device for the identification of microorganisms. *Lancet* 1960 (i), 1172–1173.

COWAN, S. T. and STEEL, K. J. (1965). "Manual for the Identification of Medical Bacteria". Cambridge University Press, Cambridge.

CROVELLO, T. J. (1968). The effect of missing data and of two sources of character values on a phenetic study of the willows of California. *Madroño* **19**, 301–315.

CROVELLO, T. J. (1970). Analysis of character variation in ecology and systematics. *A. Rev. Ecol. Syst.* **1**, 55–98.

CROVELLO, T. J. (1974). Computers in biological teaching. *BioScience* **24**, 20–23.

CUTBILL, J. L., ed. (1971a). "Data Processing in Biology and Geology". Academic Press, London and New York.

CUTBILL, J. L. (1971b). New methods for handling biological information. *Biol. J. Linn. Soc.* **3**, 253–260.

DALE, M. B. (1968). On property structure, numerical taxonomy and data handling. *In* "Modern Methods in Plant Taxonomy" (V. H. Heywood, ed.), pp. 185–197. Academic Press, London and New York.

DALLWITZ, M. J. (1974). A flexible computer program for generating identification keys. *Syst. Zool.* **23**, 50–57.

DAVIDSON, R. A. and DUNN, R. A. (1966). A new biometric approach to systematic problems, *BioScience* **16**, 528–536.

DEMIRMEN, F. (1969). Multivariate procedures and FORTRAN IV program for evaluation and improvement of classifications. *Kansas Geol. Surv. Computer Contrib.* **31**, 1–51.

DUKE, J. A. (1969). On tropical tree seedlings. I. Seeds, seedlings, systems, and systematics. *Ann. Mo. bot. Gdn* **56**, 125–161.

44   *L. E. Morse*

DYBOWSKI, W. and FRANKLIN, D. A. (1968). Conditional probability and the identification of bacteria: A pilot study. *J. gen. Microbiol.* **54,** 215–229.

ESTABROOK, G. F. and BRILL, R. C. (1969). The theory of the TAXIR accessioner. *Math. Biosci.* **5,** 327–340.

FISHER, R. A. (1936). The use of multiple measurements in taxonomic problems. *Ann. Eugen.* **7,** 179–188.

FLAKE, R. H. and TURNER, B. L. (1968). Numerical classification for taxonomic problems. *J. theor. Biol.* **20,** 260–270.

FREEMAN, H. (1974). Computer processing of line-drawing images. *Comput. Surv.* **6,** 57–97.

FURLOW, J. J. and BEAMAN, J. H. (1971). Sample taxonomic data matrices for vascular plants prepared by students at Michigan State University. *Flora N. Am. Rep.* **56,** 1–118.

GASSER, W. and GEHRT, K. M. (1971). A computer program for identifying micro-organisms. *BioScience* **21,** 1044–1045.

GERKE, J. R., HANEY, T. A., PAGANO, J. F. and FERRARI, A. (1960). Automation of the microbiological assay of antibiotics with an AutoAnalyzer instrumental system. *Ann. N.Y. Acad. Sci.* **87,** 782–791.

GERMERAAD, J. H. and MULLER, J. (1970). A computer-based numerical coding system for the description of pollen grains and spores. *Rev. Paleobot. Palynol.* **10,** 175–202.

GERMERAAD, J. H. and MULLER, J. (1974). "A Computer-based Numerical Coding System for the Description of Pollen Grains and Spores", 2 vols. Rijksmuseum van Geologie en Mineralogie, Leiden.

GILBERT, C. W, and MULDAL, S. (1971). Measurement and computer system for karyo-typing human and other cells. *Nature New Biol.* **230,** 203–207.

GILMARTIN, A. J. (1969). The quantification of some plant-taxa circumscriptions. *Am. J. Bot.* **56,** 654–663.

GILMARTIN, A. J. (1970). FNA work-in-progress on genus *Carex*. *Flora N. Am. Rep.* **31,** [61 pp.].

GILMARTIN, A. J. (1974). Variation within populations and classification. *Taxon* **23,** 523–536.

GIPSON, P. S., SEALANDER, J. A. and DUNN, J. E. (1974). The taxonomic status of wild *Canis* in Arkansas, *Syst. Zool.* **23,** 1–11.

GLASER, D. A. and WARD, C. B. (1972). Computer identification of bacteria by colony morphology. *In* "Frontiers of Pattern Recognition" (S. Watanabe, ed.), pp. 139–162. Academic Press, New York and London.

GLASER, D. A. and WATTENBERG, W. H. (1966). An automated system for the growth and analysis of large numbers of bacterial colonies using an environmental chamber and a computer-controlled flying-spot scanner. *Ann. N.Y. Acad. Sci.* **139,** 243–257.

GÓMEZ-POMPA, A. and NEVLING, L. I. Jr. (1973). The use of electronic data processing methods in the Flora of Veracruz program. *Contrib. Gray Herb.* **203,** 49–64.

GÓMEZ-POMPA, A. and SQUIRES, D. F., eds (1969). "Resúmenes del Simposio sobre Problemas de Información en las Ciencias Biológias". Inst. Biol., Univ. Nac. Autón. México, Publ. Esp. no. 1.

GOOD, I. J. (1959). Kinds of probability. *Science* **129,** 443–447.

GOODALL, D. W. (1966). Deviant index: A new tool for numerical taxonomy. *Nature, Lond.* **210,** 216.

GOODALL, D. W. (1968). Identification by computer. *BioScience* **18**, 485–488.

GOODMAN, M. M. (1967). The identification of hybrid plants in segregating populations. *Evolution* **21**, 334–340.

GORRY, G. A. and BARNETT, G. O. (1968). Experience with a model of sequential diagnosis. *Comput. Biomed. Res.* **1**, 490–507.

GOWER, J. C. and BARNETT, J. A. (1971). Selecting tests in diagnostic keys with unknown responses, *Nature, Lond.* **232**, 491–493.

GRIMES, G. M., RHOADES, H. E., ADAMS, F. C. and SCHMIDT, R. V. (1972). Identification of bacteriological unknowns: A computer-based teaching program. *J. med. Educ.* **47**, 289–292.

GUSTAFSON, D. H., EDWARDS, W., PHILLIPS, L. D. and SLACK, W. V. (1969). Subjective probabilities in medical diagnosis. *IEEE Trans. Man-Machine Syst.* **MMS-10**, 61–65.

GYLLENBERG, H. (1963). A general method for deriving determination schemes for random collections of microbial isolates. *Ann. Acad. Sci. fenn.*, ser. A, IV. Biol. **69**, 1–23.

GYLLENBERG, H. G. (1964). An approach to numerical description of microbial populations. *Ann. Acad. Sci. fenn.*, ser. A, IV. Biol. **81**, 1–23.

GYLLENBERG, H. G. (1965). A model for computer identification of micro-organisms. *J. gen. Microbiol.* **37**, 401–405.

HALL, A. V. (1965). The peculiarity index, a new function for use in numerical taxonomy. *Nature, Lond.* **206**, 952.

HALL, A. V. (1969). Group-forming and discrimination with homogeneity functions. *In* "Numerical Taxonomy" (A. J. Cole, ed.), pp. 53–68. Academic Press, London and New York.

HALL, A. V. (1970). A computer-based system for forming identification keys. *Taxon* **19**, 12–18.

HALL, A. V. (1973). The use of a computer-based system of aids for classification. *Contrib. Bolus Herb.* [Univ. Cape Town] **6**, 1–110.

HALL, R. C. and BENNETT, G. W. (1973). Pyrolysis-gas chromatography of several cockroach species. *J. Chromatogr. Sci.* **11**, 439–443.

HANKS, S. (1972). Palynotaxonomy of *Fagus* and *Nothofagus*. Ph.D. thesis, Rutgers University, New Brunswick, New Jersey.

HANSEN, B. and CUSHING, E. J. (1973). Identification of pine pollen of late Quaternary age from the Chuska Mountains, New Mexico. *Geol. Soc. Am. Bull.* **84**, 1181–1200.

HANSEN, B. and RAHN, K. (1969). Determination of angiosperm families by means of a punched-card system. *Dansk Bot. Arkiv* **26**, 1–46 + 172 punched cards.

HARRINGTON, H. D. and DURRELL, L. W. (1957). "How to Identify Plants". Swallow Press, Chicago.

HAWKES, J. G., KERSHAW, B. L. and READETT, R. C. (1968). Computer mapping of species distribution in a county flora. *Watsonia* **6**, 350–364.

HELLER, S. R. (1972). Conversational mass spectral retrieval system and its use as an aid in structure determination. *Analyt. Chem.* **44**, 1951–1961.

HESS, L. W., DUNN, D. B. and LEDICOTTE, G. (1968). Potential of activation analysis for the comparison of trace minerals in plant species. *Trans. Mo. Acad. Sci.* **2**, 91–99.

HEYWOOD, V. H. (1973). Taxonomy in crisis? *Acta bot. Acad. Sci. hung.* **19**, 139–146.

HICKS, R. R. Jr. (1973). Evaluation of morphological characters for use in identifying loblolly pine, shortleaf pine, and loblolly × shortleaf hybrids. *Castanea* **38**, 182–189.

HILL, L. R. (1974). Theoretical aspects of numerical identification. *Int. J. Syst. Bact.* **24**, 494–499.

HILL, L. R. and SILVESTRI, L. G. (1962). Quantitative methods in the systematics of the Actinomycetales. III. The taxonomic significance of physiological-biochemical characters and the construction of a diagnostic key. *Giorn. Microbiol.* **10**, 1–28.

HILLS, L. V., KLOVAN, J. E. and SWEET, A. R. (1974). *Juglans eocinerea* n. sp., Beaufort Formation (Tertiary), southwestern Banks Island, arctic Canada. *Can. J. Bot.* **52**, 65–90.

JIČÍN, R. (1972). Some problems of description of sets of objects. *J. theoret. Biol.* **34**, 295–311.

JIČÍN, R., PILOUS. Z. and VAŠÍČEK, Z. (1969). Grundlagen einer formalen methode zur Zusammenstellung und Bewertung von Bestimmungsschlüsseln. *Preslia* **41**, 71–85.

KAMP, J. W. (1973). Numerical classification of the orthopteroids, with special reference to the Grylloblattodea. *Can. Entomol.* **105**, 1235–1249.

KELLER, C. and CROVELLO, T. J. (1973). Procedures and problems in the incorporation of data from floras into a computerized data bank. *Proc. Indiana Acad. Sci.* **82**, 116–122.

KENDRICK, B. (1972). Computer graphics in fungal identification. *Can. J. Bot.* **50**, 2171–2175.

KENDRICK, W. B. (1965). Complexity and dependence in computer taxonomy. *Taxon* **14**, 141–154.

KIM, K. C., BROWN, B. W. Jr. and COOK, E. F. (1966). A quantitative taxonomic study of the *Hoplopleura hesperomydis* complex (Anoplura, Hoplopleuridae), with notes on *a posteriori* taxonomic characters. *Syst. Zool.* **15**, 24–45. ,

KING, W. B., WATSON, G. E. and GOULD, P. J. (1967). An application of automatic data processing to the study of seabirds. I. Numerical coding. *Proc. U.S. natn. Mus.* **123**, no. 3609, 1–29.

KNUTH, D. E. (1973). Computer science and mathematics. *Am. Scient.* **61**, 707–713.

KRAUSS, H. M. (1973). The use of generalized information processing systems in the biological sciences. *Taxon* **22**, 3–18.

KRICHEVSKY, M. I. and NORTON, L. M. (1974). Storage and manipulation of data by computers for determinative bacteriology. *Int. J. Syst. Bact.* **24**, 524–531.

LANDOWNE, R. A., MOROSANI, R. W., HERMANN, R. A., KING, R. M. Jr. and SCHMUS, H. G. (1972). Computer acquisition and analysis of gas chromatographic data. *Analyt. Chem.* **44**, 1961–1971.

LAPAGE, S. P. (1974). Practical aspects of probabilistic identification of bacteria. *Int. J. Syst. Bact.* **24**, 500–507.

LAPAGE, S. P. and BASCOMB, S. (1968). Use of selenite reduction in bacterial classification. *J. appl. Bacteriol.* **31**, 568–580.

LAPAGE, S. P., BASCOMB, S., WILLCOX, W. R. and CURTIS, M. A. (1970). Computer identification of bacteria. *In* "Automation, Mechanization, and Data Handling in Microbiology" (A. Baillie and R. J. Gilbert, eds), pp. 1–22. Academic Press, London and New York.

LAPAGE, S. P., BASCOMB, S., WILLCOX, W. R. and CURTIS, M. A. (1973). Identification of bacteria by computer: General aspects and perspectives. *J. gen. Microbiol.* **77**, 273–290.

LAWRENCE, B. and BOSSERT, W. H. (1967). Multiple character analysis of *Canis lupis, latrans* and *familiaris*, with a discussion of the relationships of *Canis niger*. *Am. Zool.* **7**, 223–232.

LEDLEY, R. S. (1969). Practical problems in the use of computers in medical diagnosis. Proc. IEEE [Inst. Electrical and Electronics Eng.] **57,** 1900–1918.

LEDLEY, R. S. and LUSTED, L. B. (1959a). Reasoning foundations of medical diagnosis. *Science* **130,** 9–21.

LEDLEY, R. S. and LUSTED, L. B. (1959b). The use of electronic digital computers to aid in medical diagnosis. *Proc. Inst. Radio Eng.* **47,** 1970–1977.

LEDLEY, R. S. and RUDDLE, F. H. (1966). Chromosome analysis by computer. *Scient. Am.* **214,** no. 4, 40–46.

LEENHOUTS, P. W. (1966). Keys in biology: A survey and a proposal of a new kind. *Proc. K. ned. Akad. Wet.* **69C,** 571–596.

LITTLE, E. L. Jr. (1968). Clave con fichas perforadas de las familias de los árboles mexicanos. *Turrialba* **18,** 45–59.

LUSTED, L. B. (1968). "Introduction to Medical Decision-making". Charles C. Thomas, Springfield, Ill.

MACCACARO, G. A. (1958). La misura della informazione contenuta nei criteri di classificazione. *Ann. Microbiol. Enzimol.* **8,** 231–239.

MACHOL, R. E. and SINGER, R. (1971). Bayesian analysis of generic relations in Agaricales. *Nova Hedwigia* **21,** 753–787.

MACINNES, J. R., RHODES, E. W. and CALABRESE, A. (1974). A new electronic system for counting and measuring bivalve larvae. *Chesapeake Sci.* **15,** 174–176.

MARTIN, J. (1973). "Design of Man–Computer Dialogues". Prentice-Hall, Englewood Cliffs, New Jersey.

MAYR, E. (1965). Numerical phenetics and taxonomic theory. *Syst. Zool.* **14,** 73–97.

MAYR, E. (1969). "Principles of Systematic Zoology". McGraw-Hill, New York.

MAYR, E. (1971). Methods and strategies in taxonomic research. *Syst. Zool.* **20,** 426–433.

McHENRY, H. M. (1973). Early Hominid humerus from East Rudolf, Kenya. *Science* **180,** 739–741.

McNEILL, J. (1972). The hierarchical ordering of characters as a solution to the dependent character problem in numerical taxonomy. *Taxon* **21,** 71–82.

MEIJER, W. (1967). Materials toward a foresters' flora of Sabah. *In* "Natural Resources in Malaysia and Singapore" (B. C. Stone, ed.), pp. 17–23. 2nd. Symp. Sci. Techn. Res. Mal. & Singapore.

MENGER, F. M., EPSTEIN, G. A., GOLDGERG, D. A. and REINER, E. (1972). Computer matching of pyrolysis chromatograms of pathenogenic microorganisms. *Analyt. Chem.* **44,** 423–424.

METCALF, Z. P. (1954). The construction of keys. *Syst. Zool.* **3,** 38–45.

MEUZELAAR, H. L. C. (1974). "Identification of bacteria by pyrolysis gas chromatography and pyrolysis mass spectrometry". Academisch proefschrift, Univ. Amsterdam.

MEUZELAAR, H. L. C. and KISTEMAKER, P. G. (1973). A technique for fast and reproducible fingerprinting of bacteria by pyrolysis mass spectroscopy. *Analyt. Chem.* **45,** 587–590.

MICHENER, C. D. *et al.* (1970). "Systematics in Support of Biological Research". Div. Biol. Agric., Natn. Res. Council, Washington, D.C.

MÖLLER, F. (1962). Quantitative methods in the systematics of the Actinomycetales. IV. The theory and application of a probabilistic identification key. *Giorn. Microbiol.* **10,** 29–47.

MORSE, L. E. (1968). Construction of identification keys by computer (Abstract). *Am. J. Bot.* **55**, 737.

MORSE, L. E. (1969). Time-sharing computers as aids to identification of plant specimens (Demonstration). *Abstr. XI Int. Bot. Congr.*, 152.

MORSE, L. E. (1970). Computer aids to plant identification (Abstract). *Am. J. Bot.* **57**, 754–755.

MORSE. L. E. (1971). Specimen identification and key construction with time-sharing computers. *Taxon* **20**, 269–282.

MORSE, L. E. (1974a). Computer programs for specimen identification, key construction, and description printing using taxonomic data matrices. *Publs Mus. Michigan State Univ., Biol. Ser.* **5**, 1–128.

MORSE, L. E. (1974b). Computer-assisted storage and retrieval of the data of taxonomy and systematics. *Taxon* **23**, 29–43.

MORSE, L. E., BEAMAN, J. H. and SHETLER, S. G. (1968). A computer system for editing diagnostic keys for Flora North America. *Taxon* **17**, 479–483.

MORSE, L. E., BEAMAN, J. H. and SHETLER, S. G. (1969). Preparation of identification keys by computer for Flora North America (Abstract). *Inst. Biol., Univ. Nac. Autón. México, Publ. Esp.* **1**, 22.

MORSE, L. E., PETERS, J. A. and HAMEL, P. B. (1971). A general data format for summarizing taxonomic information. *BioScience* **21**, 174–180, 186.

MULLIN, J. K. (1970). COQAB: A computer optimized question asker for bacteriological specimen identification. *Math. Biosci.* **6**, 55–66.

MYERS, A. and WATSON, L. (1969). Rapid diagnosis of viral and fungal diseases in plants by pyrolysis and gas–liquid chromatography. *Nature, Lond.* **223**, 964–965.

NEWELL, I. M. (1970). Construction and use of tabular keys. *Pacific Insects* **12**, 25–37.

NICHOLS, H. W., ANDERSON, D. J., SHAW, J. I. and SOMMERFIELD, M. R. (1968). Pyrolysis–gas–liquid chromatographic analysis of chlorophycean and rhodophycean algae. *J. Phycol.* **4**, 362–368.

NIEMELÄ, S. I., HOPKINS, J. W. and QUADLING, C. (1968). Selecting an economical binary test battery for a set of microbial cultures. *Can. J. Microbiol.* **14**, 271–279.

NIEVERGELT, J. (1974). Binary search trees and file organization. *Comput. Surv.* **6**, 195–207.

NORMAND, D. (1946). Les clés pour l'identification des bois et le systeme des fiches perforées. *Agron. trop.* **1**, 162–172.

OGDEN, E. C. (1953). Key to the North American species of *Potamogeton. N.Y. State Mus., Circ.* **31**, 1–11.

OLDS, R. J. (1970). Identification of bacteria with the aid of an inproved imformation sorter. *In* "Automation, Mechanization, and Data Handling in Microbiology" (A. Baillie and R. J. Gilbert, eds), pp. 85–89. Academic Press, London and New York.

OSBORNE, D. V. (1963a). A numerical representation for taxonomic keys. *New Phytol.* **62**, 35–43.

OSBORNE, D. V. (1963b). Some aspects of the theory of dichotomous keys. *New Phytol.* **62**, 144–160.

PANKHURST, R. J. (1970a). A computer program for generating diagnostic keys. *Comput. J.* **13**, 145–151.

PANKHURST, R. J. (1970b). Key generation by computer. *Nature, Lond.* **227**, 1269–1270.

PANKHURST, R. J. (1971). Botanical keys generated by computer. *Watsonia* **8**, 357–368.

PANKHURST, R. J. (1972). A method for data capture. *Taxon* **21**, 549–558.

PANKHURST, R. J. (1974). Automated identification in systematics. *Taxon* **23**, 45–51.

PANKHURST, R. J. and WALTERS, S. M. (1971). Generation of keys by computer. *In* "Data Processing in Biology and Geology" (J. L. Cutbill. ed.), pp. 189–203. Academic Press, London and New York.

PETERS, J. A. (1968). A computer program for calculating degree of biogeographical resemblance between areas. *Syst. Zool.* **17**, 64–69.

PETERS, J. A. (1969a). Computer techniques in systematics: Discussion. *In* "Systematic Biology". pp. 610–613, publ. 1692, National Academy of Sciences, Washington, D.C.

PETERS, J. A. (1969b). Preparación y manipulación de claves sistemáticas utilizando computadoras de tiempo compartido (Abstract). *Inst. Biol., Univ. Nac. Autón. México.*, Publ. Esp. **1**, 25.

PETERS, J. A. (1972). Museum and university data, program, and information exchange. *MUDPIE* [Mus. Univ. Data Prog. Info. Exch., Smithsonian Inst.] no. 26, 2–6.

PETERS, J. A. and COLLETTE, B. B. (1968). The role of time-share computing in museum research. *Curator* **11**, 65–75.

PETTIGREW, C. J., and WATSON, L. (1973). On the identification of sterile acacias, and the feasibility of establishing an automatic key-generating system. *Aust. J. Bot.* **21**, 141–150.

PIGUET, J. D. and ROBERGE, P. (1970). Problèmes posés par le diagnostic automatique des bâtonnets Gram-négatifs. *Can. J. publ. Hlth* **61**, 329–335.

PORTER, D. M., KIGER, R. W. and MONAHAN, J. E. (1973). A guide for contributors to Flora North America, Part II. An outline and glossary of terms for morphological and habitat description (Provisional ed.). *Flora N. Am. Rep.* **66**.

PREECE, T. F. (1968). The fluorescent antibody technique for the identification of plant pathogenic fungi. *In* "Chemotaxonomy and Serotaxonomy (J. G. Hawkes. ed.), pp. 111–114. Academic Press, London and New York.

REINER, E. (1965). Identification of bacterial strains by pyrolysis–gas–liquid chromatography. *Nature, Lond.* **206**, 1272–1274.

REINER, E. and EWING, W. H. (1968). Chemotaxonomic studies of some Gram-negative bacteria by means of pyrolysis–gas–liquid chromatography. *Nature, Lond.* **217**, 191–194.

RIDDLE, J. W., KABER, P. W., KENNER, B. I., BORDNER, R. H., ROCKWOOD, S. W. and STEVENSON, H. J. R. (1956). Bacterial identification by infrared spectrophotometry. *J. Bact.* **72**, 593–603.

ROGERS, D. J. (1970). Theoretical and practical considerations on data structuring for a computerized information retrieval system. *In* "Archéologie et Calculateurs", pp. 145–159. Centre National de la Recherche Scientifique, Paris.

ROGERS, D. J. and FLEMING, H. S. (1973). A monograph of *Manihot esculenta*—with an explanation of the taximetric methods used. *Econ. Bot.* **27**, 1–113.

ROHLF, F. J. and SOKAL, R. R. (1967). Taxonomic structure from randomly and systematically scanned biological images. *Syst. Zool.* **16**, 246–260.

ROLLINS, R. C. (1965). On the bases of biological classification. *Taxon* **14**, 1–6.

ROLLINS, R. C. and SOLBRIG, O. T. (1973). Interspecific hybridization in *Lesquerella*. *Contr. Gray Herb.* **203**, 3–48.

ROSENBERG, S. A., LEDEEN, K. S. and KLINE, T. (1969). Automatic identification and measurement of cells by computer. *Science* **163**, 1065–1066.

ROSENFELD, A. (1973). Progress in picture processing: 1969–71. *Comput. Surv.* **5**, 81–108.

RUNEMARK, H. (1968). Critical comments on the use of statistical methods in chemotaxonomy. *Bot. Notiser* **121**, 29–43.

RUTOVITZ, D. (1968). Automatic chromosome analysis. *Br. med. Bull.* **24**, 260–267.

RYPKA, E. W. (1971). Truth table classification and identification. *Space Life Sci.* **3**, 135–156.

RYPKA, E. W. and BABB, R. (1970). Automatic construction and use of an identification scheme. *Med. Res. Engng.* **9**, 9–19.

RYPKA, E. W., CLAPPER, W. E., BOWEN, I. G. and BABB, R. (1967). A model for the identification of bacteria. *J. gen. Microbiol.* **46**, 407–424.

SAILA, S. B. (1970). [Review of Pankhurst's "A computer program for generating diagnostic keys"] *Comput. Rev.* **11**, 545.

SEVIOUR, R. J., CHILVERS, G. A. and CROW, W. D. (1974). Characterization of eucalypt mycorrhizas by pyrolysis–gas chromatography. *New Phytol.* **73**, 321–332.

SHETLER, S. G. (1971). Flora North America as an information system. *BioScience* **21**, 524, 529–532.

SHETLER, S. G. (1974). Demythologizing biological data banking. *Taxon* **23**, 71–100.

SHETLER, S. G., BEAMAN, J. H., HALE, M. E., MORSE, L. E., CROCKETT, J. J. and CREIGHTON, R. A. (1971). Pilot data processing systems for floristic information. *In* "Data Processing in Biology and Geology" (J. L. Cutbill, ed.), pp. 275–310. Academic Press, London and New York.

SHULTZ, L. M. (1973). "Random-access Key to Genera of Colorado Wildflowers". University of Colorado Museum, Boulder.

SNEATH, P. H. A. (1957). Some thoughts on bacterial classification. *J. gen. Microbiol.* **17**, 184–200.

SNEATH, P. H. A. (1962). The construction of taxonomic groups. *Symp. Soc. gen. Microbiol.* **12**, 289–332.

SNEATH, P. H. A. (1969). Computers in bacteriology. *J. clin. Pathol.* **22**, suppl. (College of Pathologists), no. 3, 87–92.

SNEATH, P. H. A. (1971). Numerical taxonomy: Criticisms and critiques. *Biol. J. Linn. Soc.* **3**, 147–157.

SNEATH, P. H. A. (1974). Test reproducibility in relation to identification. *Int. J. Syst. Bact.* **24**, 508–523.

SNEATH, P. H. A. and COWAN, S. T. (1958). An electro-taxonomic survey of bacteria. *J. gen. Microbiol.* **19**, 551–565.

SNEATH, P. H. A. and JOHNSON, R. (1972). The influence on numerical taxonomic similarities of errors in microbiological tests. *J. gen. Microbiol.* **72**, 377–392.

SNEATH, P. H. A. and SOKAL, R. R. (1973). "Numerical Taxonomy: The Principles and Practice of Numerical Classification". Freeman, San Francisco.

SOKAL, R. R. (1965). Statistical methods in systematics. *Biol. Rev.* **40**, 337–391.

SOKAL, R. R. and SNEATH, P. H. A. (1963). "Principles of Numerical Taxonomy". Freeman, San Francisco and London.

SOKAL, R. R. and SNEATH, P. H. A. (1966). Efficiency in taxonomy. *Taxon* **15**, 1–21.

SOLBRIG, O. T. (1970). "Principles and Methods of Plant Biosystematics". Macmillan, New York.

SPRUNG, D. C. and WUJEK, D. E. (1974). Chemotaxonomic studies of *Pleurastrum* Chodat by means of pyrolysis–gas–liquid chromatography. *Phycologia* **10**, 251–254.

STEEL, K. J. (1962). The practice of bacterial identification. *Symp. Soc. gen. Microbiol.* **12,** 405–432.

STEPHENSON, S. N. (1971). The biosystematics and ecology of the genus *Brachyelytrum* (Gramineae) in Michigan. *Michigan Bot.* **10,** 19–33.

STEVENS, W. D. (1971). The general taxonomic data matrix format used for *Sarcostemma* (Asclepiadaceae). *Flora N. Am. Rep.* **63,** 43–66.

TAGGART, R. E. (1971). Automation of identification procedures in palynology using taxonomic data matrices. *Flora N. Am. Rep.* **63,** 67–90.

TUCKER, J. M. (1968). Identity of the oak tree at Live Oak Tank, Joshua Tree National Monument, California. *Madroño* **19,** 256–266.

TURRILL, W. B. (1957). The subjective element in plant taxonomy. *Bull. Jard. bot. Brux.* **27,** 1–8.

VAN DAM, A. and RICE, D. E. (1971). On-line text editing: a survey. *Comput. Surv.* **3,** 93–114.

VINCENT, P. G. and KULIK, M. M. (1970). Pyrolysis–gas–liquid chromatography of fungi: Differentiation of species and strains of several members of the *Aspergillus flavus* group. *Appl. Microbiol.* **20,** 957–963.

VINCENT, P. G. and KULIK, M. M. (1973). Pyrolysis–gas–liquid chromatography of fungi: Numerical characterization of species variation among members of the *Aspergillus glaucus* group. *Mycopathol. Mycol. appl.* **51,** 251–265.

VLCEK, J. (1972). "Considerations in determination, evaluation and computer banking of spectral signatures of natural objects". Internal Rep. FMR-22, Forest Management Institute, Ottawa.

VOGT, W. G. and MCPHERSON, D. G. (1972). The weighted separation index: a multivariate technique for separating members of closely-related species using qualitative differences. *Syst. Zool.* **21,** 187–198.

VOSS, E. G. (1952). The history of keys and phylogenetic trees in systematic biology. *J. Sci. Lab., Denison Univ.* **43,** 1–25.

VOSS, E. G. (1972). "Michigan Flora, Part I. Gymnosperms and Monocots". Cranbrook Institute of Science, Bloomfield Hills, Mich., and University of Michigan Herbarium, Ann Arbor.

WAGNER, W. H. Jr. (1954). Reticulate evolution in the Appalachian Aspleniums. *Evolution* **8,** 103–118.

WAGNER, W. H. Jr. (1969). The Pteridophyta in the Flora of North America: The role and taxonomic treatment of hybrids. *BioScience* **19,** 785–787.

WALKER, D., MILNE, P., GUPPY, J. and WILLIAMS, J. (1968). The computer assisted storage and retrieval of pollen morphological data. *Pollen et Spores* **10,** 251–262.

WATSON, L. (1971). Basic taxonomic data: The need for organization over presentation and accumulation. *Taxon* **20,** 131–136.

WATSON, L. and MILNE, P. (1972). A flexible system for automatic generation of special purpose dichotomous keys, and its application to Australian grass genera. *Aust. J. Bot.* **20,** 331–352.

WEBB, L. J., TRACEY, J. G., WILLIAMS, W. T. and LANCE, G. N. (1967). Studies in the numerical analysis of complex rain-forest communities. I. A comparison of methods applicable to site/species data. *J. Ecol.* **55,** 171–191.

WEBER, W. A. and NELSON, P. P. (1972). "Random-access key to genera of Colorado mosses". University of Colorado Museum, Boulder.

WEBSTER, T. (1969). Developments in the description of potato varieties, Part I—Foliage. *J. natn. Inst. agric. Bot.* **11**(3), 455–475.

WEBSTER, T. (1970). Developments in the description of potato varieties, Part II—Inflorescences and tubers. *J. natn. Inst. agric. Bot.* **12**(1), 17–45.

WHIFFIN, T. (1973). Analysis of a hybrid swarm between *Heterocentron elegans* and *H. glandulosum* (Melastomataceae). *Taxon* **22**, 413–423.

WHITEHEAD, F. H. (1954). An example of discrimination by biometric methods. *New Phytol.* **53**, 496–510.

WHITEHEAD, P. P. J. (1971). Storage and retrieval of information in systematic zoology. *Biol. J. Linn. Soc.* **3**, 211–220.

WILLCOX, W. R. and LAPAGE, S. P. (1972). Automatic construction of diagnostic tables. *Comput. J.* **15**, 263–267.

WILLCOX, W. R., LAPAGE, S. P., BASCOMB, S. and CURTIS, M. A. (1973). Identification of bacteria by computer: Theory and programming. *J. gen. Microbiol.* **77**, 317–330.

WILLIAMS, W. T. (1967). The computer botanist. *Aust. J. Sci.* **29**, 266–271.

WILLIAMS, W. T. (1969). The problem of attribute-weighting in numerical classification. *Taxon* **18**, 369–374.

ZADEH, L. A. (1965). Fuzzy Sets. *Info. Control* **8**, 338–353.

Techniques

3 | A System for Automatic Key-forming

A. V. HALL

Bolus Herbarium, University of Cape Town, Republic of South Africa

Abstract: The automatic production of dichotomous identification keys, by means of a computer program and numerical data, is discussed. The working environments in which such systems are likely to be useful are specified. A key-forming system which is part of a program-package of classification aids is described. Practical aspects of the system are discussed and the problem of scanning ahead to show the consequences of choosing a particular test at a dichotomy is noted. It is concluded that the automatic production of dichotomous keys may indeed become a useful aid to the taxonomist.

Key Words and Phrases: dichotomous keys, taxonomic data matrix, polyclaves, modal data, discrete data, grouping

INTRODUCTION

A computer-based system, for producing dichotomous keys for identifying biological specimens, has been revised and extended (Hall, 1970, 1973). The system is designed to be efficient in a specific working environment. The aim of this paper is to show how the worker's needs are catered for by the key-forming system, within this environment.

MAKING IDENTIFICATION KEYS

Most keys are made by monographers or the authors of regional guides to Floras and Faunas. In large monographs, close relationships and many taxa may make the construction of an efficient key a somewhat difficult task. The initial tests may require very detailed study for their preparation, in order to provide evenly balanced dichotomies that will work with minimum chance of error. The monographer is well equipped to make such a key: he usually comes to it at the end of a thorough study of the species making up his group. However, the monographer's approach is very broadly based, with a study of all available properties from taxa from perhaps many regions. The key that he writes may

Systematics Association Special Volume No. 7, "Biological Identification with Computers", edited by R. J. Pankhurst, 1975, pp. 55–63. Academic Press, London and New York.

see rather less use than the keys drawn up by other workers for local areas, some adapted for special classes of characters, for cases such as plants in their non-flowering or leafless states. The monographer has access to a great deal of information that would be of use in these local projects. Through the sheer volume of the data, much of it remains unrecorded by the monographer. The writer of the local "Flora" or "Fauna" mostly has to start afresh in adding to his knowledge of the taxa before making keys that he may consider to be efficient.

The writer of the local Flora or Fauna generally has fewer examples of each major taxon than the monographer. Although his work is easier in this respect, there is the difficult practical problem of making keys for all chief growth-forms of the local organisms throughout the year. In available Floras and Faunas, how often can we say this is thoroughly done? In a regional work, would it be desirable to have extra keys in large, difficult groups, each restricted to important local areas or habitats? In the very rich Cape Flora at the southern tip of Africa, the special problem of plant identification is greatly eased for the specialist when a locality or habitat is given.

These problems are important. Identification procedures are the route for relating biological work to the standard namings laid down by the taxonomist. Efficiency along this route is as critically important for allied disciplines, such as environmental studies and agriculture, as for biology itself. How can maximum efficiency be achieved?

SPECIFICATIONS FOR THE DEVELOPMENT OF EFFICIENT KEYS

The above discussion suggests that the best keys are those that are adapted to the needs of the moment: explicitly, to some subset of characters and to the organisms known from a local area or a special habitat. A very good way of achieving this is a polyclave key, consisting of "peep-hole" cards, punched at set positions for the taxa having a particular character-state, or habitat, or localized distribution. The cards can be used in any combination to get to an answer, giving the polyclave very great powers of adaptability (this volume: Pankhurst, Chapter 5; Morse, p. 25; Leenhouts, 1966). The division of subcontinous data into classes, which are the "character states" for many of the cards, causes some loss of information. This loss is probably well compensated by the user being able to employ many characters together. The boundaries of the classes have to be chosen with care so as to make maximum use of the discriminative ability of the properties. In very difficult groups many cards may have to be used together to separate off closely allied taxa. An important point is that the polyclave, like the written dichotomous key, does not provide a confirming description or illustration. This confirmation is an essential part of the identification process, especially in borderline cases. Identifications should not be carried out with a deck of polyclave cards alone.

An alternative to the polyclave is to use a computer-based key-forming algorithm that makes maximum use of the data to produce a printed, dichotomous key. This allows the following needs of the working environment for key-forming to be met:

(1) The system sets up the best compromise for identification routes for all taxa, following accurately an information-rich, numerical data base.

(2) Once the data have been collected, they can be used for the automatic production of keys for any subset of taxa, emphasizing some properties and perhaps omitting others.

(3) Optimization procedures, for making all the keys as efficient as possible, can be written into the computer program: tests are chosen with the fewest characters, with the most easily observable properties and with equal likely usage of the branches of each dichotomy; also, the most common taxa can be made to appear early in the key (Hall, 1970; Pankhurst, 1970).

The computer-formed dichotomous key has to be made from data that give a numerical image of the taxa. Setting down the list of properties and measuring, classing and scoring them (as departures from basic "ground-plan" states), can be a long job. A similar problem exists for making polyclaves. However, if the numerical data are recorded for a full set of properties, covering all aspects of the organisms, they can also be used for forming groups, showing group cores and edges, describing trends among the taxa and other operations, all of which can indeed be of great value in monographic studies (Hall, 1965). It would seem then that the effort of numerical coding and using the computer for key-forming may at present be justified in the following cases:

(1) where the monographer is working with a large and difficult taxon, in which both key-forming and the analysis of group structure would benefit by computer aids;

(2) in cases of larger sets of taxa in a regional guide to a Flora or Fauna, in which several keys are needed for special areas or habitats, or where only limited subsets of properties can be used.

In the future, it may well be profitable to extend the use of automatic key-forming, in some regions, to all groups so that special identification manuals can be issued, fully corrected and up to date, in a highly automated way. A difficulty in achieving this is the problem of processing large amounts of verbal data. However, both Morse (1974) and Pankhurst (1972) have made important contributions in this respect.

C

THE BOLAID KEY-FORMING SYSTEM
THE BOLAID KEY-FORMING SYSTEM
A system for making dichotomous keys was included in a general program-package of classification aids, with the above considerations in mind (BOLAID: Hall, 1973).

1. Input

The BOLAID package uses numerical data alone; small changes in the input instructions can be made to cause one of a variety of operations to be carried out. The data can be presented in two-state, multi-state or sub-continuous (i.e. quantitative) form. As the program uses numerical and not verbal descriptions of taxa, the results given in the printout have to be rewritten as full statements, as shown in Figs 1 and 2. This is a rapid and simple operation. Full editing and proof-reading facilities for the numerical data are given in the package.

An important aspect is that ranges to show the variation of quantitative properties can be given. This is an almost essential feature of a key-forming system. Ranges are often a source of difficulty when trying to construct keys by eye, and the user will appreciate the advantage of being able to specify them for accurate inspection by a computer program.

Other inputs to the program include calling for keys using subsets of the list of properties, or excluding some of the taxa. This permits the automatic production of keys for special local needs, or avoiding say the use of floral characters for an "out-of-season" key.

Properties that are hard to observe are flagged for de-emphasis in the computer's construction of the key. This allows the production of keys for the very diverse observing facilities of the laboratory, with preserved or fresh material, as contrasted with field conditions.

Some taxa may need to be keyed out much more often than others. An important input to the program is a list of expected key-usage frequencies for the taxa. The values may be highest for the not-so-common taxa, less if they are very well known and least if they are rare.

2. Treatment of the data

The program is written to find the best property or pair of properties for dividing a set of taxa into two groups. Further divisions are run successively until all sets have been divided down to individual taxa. Some groups may be indivisible, necessitating the use of further data. The program adds an accessory character to the chosen property or property-pair, to give further guidance to the user (see Fig. 1). For the purposes of the system, properties are regarded as either: (a) non-modal: with the data in discrete states with no imaginable raw values in between them; or (b) modal: quantitative, with actual or potential intermediate modes or values.

The testing of non-modal data is rather straightforward, as it is assumed that there is a sharp distinction between any pair of states. Genuine non-modal

```
1A CHARACTER      9, RANGE        2 -     40

    2A CHARACTER      9, RANGE     40  -     40
       ACCESSORY      8, RANGE    100  -    100
                  .......... ITEM    5

    2B CHARACTER      9, RANGE      2  -      5
       ACCESSORY      8, RANGE      0  -      0

        3A CHARACTER      7, RANGE     22  -     22
                      .......... ITEM    1

        3B CHARACTER      7, RANGE     30  -     38

            4A CHARACTER     11, RANGE     10  -     10
                          .......... ITEM    6
            4B CHARACTER     11, RANGE     30  -     30
                          .......... ITEM    2

1B CHARACTER      9, RANGE       70  -    100

    5A CHARACTER      7, RANGE      22  -     22
       ACCESSORY     11, RANGE       8  -      8
                  ........... ITEM    3

    5B CHARACTER      7, RANGE      42  -     52
       ACCESSORY     11, RANGE      10  -     12

        6A CHARACTER     10, RANGE       7  -     11
           CHARACTER      5, RANGE      20  -     20
                      ........... ITEM    7
        6B CHARACTER     10, RANGE      16  -     22
           CHARACTER      5, RANGE       0  -      0
                      ........... ITEM    4
```

FIG. 1. Showing the computer print-out for an optimized key to six species of the Orchid genus *Satyrium* Sw., showing an evenly-balanced first dichotomy. Only modal data were used, several without details of their ranges within each species.

properties are rare in biological work, and some relaxation of the above definition may be used in key-forming, for cases where the organisms show clearly distinct states.

Modal data are much more difficult to process, requiring a special classing system, as follows. For every subset being divided, the program scales the values of modal properties to a range from zero to 20. This allows standard comparisons to be made among the properties. The occurrences of values in 21 classes along the property range are found. This is done for single properties and all possible pairs. The pairs are tested for mutual reinforcement of discriminative ability, with both classed in the same sequence of magnitudes, as well as the case with one property classed in the opposite "direction". These more complex relations are not examined if at least five very clear distinctions have been found using easily observed single properties.

1a. Perianth green to pink.

   2a. Perianth pink, lip flap hairy .... *S. longicauda* (5)

   2b. Perianth green, lip flap not hairy.

      3a. Flowers small, galea 3 mm high .... *S. parviflorum* (1)

      3b. Flowers medium-sized, galea 5–6 mm high.

         4a. Stigma as long as broad .... *S. stenopetalum* (6)

         4b. Stigma three times as long as broad .... *S. odorum* (2)

1b. Perianth orange to deep red.

   5a. Flowers small, lip galea 3 mm high .... *S. neglectum* (3)

   5b. Flowers large, lip galea 7–9 mm high.

      6a. Spurs 7–11 mm long, leaves spotted below ...... *S. coriifolium* (7)

      6b. Spurs 16–22 mm long, leaves not spotted   ...... *S. woodii* (4)

FIG. 2. Showing the full key written from the results given in Fig. 1. Species with the highest key-usage frequencies are given first in the key: taxa 5 and 1, both having the largest value (100) are placed first, while *Satyrium woodii* with the lowest (10), comes last.

For each property or property-combination, the classing system is inspected by the program in a search for the best possible gap that will allow the taxa to be split into two sets. This is done with an array that records the number of values or ranges that lie across sets of classes. Once a gap is found, its midpoint is located and its importance is assessed by the distribution of values and ranges in the neighbourhood (Hall, 1970).

At first gaps are only sought within a narrow range, set to give a "well-balanced" dichotomy. "Balance" is here based on the likely usage of each branch, acquired from the item-frequencies given with the input for the run. Subsequent passes through the properties are made with wider ranges of acceptance for the site of the gap. Every value expressing the dispersion of ariation at a taxon/property intersection in the data matrix results in a frequency entry being added at the corresponding class. If five values express the range and dispersion of a variable property of a taxon, all will be reflected in the classing system that guides the formation of the key. The function that gives the level of equality of use of the two subsets is corrected to allow the balance criterion to have little effect on small values: it is only in larger groups that the inefficiency of "lopsided" key structures becomes evident (Osborne, 1963). The proportional limits permitted at the three acceptance levels of the program are given in Table I.

TABLE I. The limits of usage-frequencies employed by the BOLAID key-forming system for the branches of dichotomies. The limits are given for the three acceptance thresholds of the program

Set size or total usage frequency	4	8	30	200
Acceptance threshold				
0·8	1–3	3–5	12–18	80–120
0·6	1–3	2–6	9–21	60–140
0·00001	1–3	1–7	1–29	1–199

Optimization for the use of the most easily observable properties is carried out in three ways. Firstly, the properties are set in a sequence according to the input ease-of-observation values. This ensures that for every trial in the program, the most conspicuous properties receive prior attention. Secondly, acceptance thresholds are set in the program which allow consideration of properties only with ease-of-observation values above certain levels. The thresholds are reduced for each subsequent pass, as with the balance limits above. Finally, ease-of-observation is allowed to play a major part in a function for comparing the performance of one property against another. This function also uses balance data, as well as a measure of the width of the separating gap in modal properties.

As soon as one or two properties have been chosen for a division, the program searches for an additional character which, perhaps being less clear-cut, will nevertheless give valuable extra guidance for the user. The aim is to find the property which lies above the ease-of-observation limit for the current pass, and which gives the clearest separation of the two new subsets of taxa.

3. Output
The print-out is somewhat simplified as a result of programming considerations (see Fig. 1). Indentations are given for sub-ordinate couplets, and states and

ranges are provided for the numbered characters. The results are set down to give the most commonly identified taxa first. In the absence of usage-frequencies, the larger group at a division is keyed out before the other. These arrangements allow the taxonomist to write out the key directly. Importantly, in composing the statement for each lead in the key, the taxonomist may be able to draw upon his specialized experience with the group. In this respect, the key-forming system is seen as an aid rather than an attempt at taking the place of a highly skilled activity.

4. *Data preparation and computing times*

The program is at present written to accept a total of 10 000 values in the raw data matrix. This limit may be expanded in the direction of the programmed maxima of 500 taxa and 999 strings of property data. Preparing a data matrix for 200 items and 50 properties would take close to four 40 h weeks at a rate of one entry per minute. Faster rates are likely with most properties.

Computing times vary with the size of the data matrix, the number of trials needed before a suitable character is found, and whether mostly modal data are used. With the UNIVAC 1106 computer, 17 taxa and 14 properties, two with extra data strings to show variation, were processed to give a key in 11 s. Computing times should be substantial for large sets of complex modal data that do not readily give balanced dichotomies. However, it would seem that the improvements in the computing speeds of hardware expected in the coming years will make this a matter of comparatively small importance.

DISCUSSION AND CONCLUSIONS

The system described above would certainly seem to be workable as an aid in key-forming. The program operates in an information-rich way, with enough data to give the essentials of a highly optimized key.

However, one of the most difficult aspects of key-forming is considering, at each step, the later consequences of choosing one dichotomy against others: will the resulting subsets be easily divisible? Would some other subsets be broken down more easily? In a rather difficult key to a group of 40 South African orchid taxa, two apparently good choices of characters had to be abandoned for this reason (Hall, 1965). Scanning ahead, to show the consequences of adopting each respective alternative at a division, would lead to absurdly large computing times if run for every major subset. It would seem best to have the user choose the sites at which alternative routes are to be explored. These would be indicated by failure to resolve a subset, or the use of hard-to-observe characters, or the presence of a severely unbalanced key. Regulating the re-runs from a terminal would seem to be ideal for this. In the present program, the subset giving difficulty can be isolated for separate runs, with the initially chosen characters de-weighted to near or below the ease-of-observation threshold at which characters are usable, currently 20 on a 0–100 scale.

A practical problem that may be unimportant for most studies, is that the data indicating the distributions of variable properties for grouping studies may have to be changed for key-forming. This is because the outermost of the values showing the dispersion of a frequency distribution may not be at the limits of the range. Frequency distributions are needed for group-forming, while ranges must be given for making keys. In practice, the difference may not be serious, especially in view of the fact that the key-forming system needs at least a significant gap between class-ranges to give separation of the groups.

Experience with the BOLAID system has shown that a desirable addition to the print-out would be some indication of the frequency distribution of values at a division. This allows a useful guide in drawing up the verbal statements for the final key (Hall, 1970).

In conclusion, it would seem that provided the rather tedious process of making a numerical image of the taxa can be justified, the computer-based production of dichotomous keys may indeed be a useful aid to the taxonomist.

REFERENCES

HALL, A. V. (1965). Studies of the South African species of *Eulophia*. *S.Afr. J. Bot.* **5** (Suppl.), 1–248.

HALL, A. V. (1970). A computer-based system for forming identification keys. *Taxon* **19**, 12–18.

HALL, A. V. (1973). The use of a computer-based system of aids for classification. *Contr. Bolus Herb.* **6**, 1–110.

LEENHOUTS, P. W. (1966). Keys in Biology. II. A survey and a proposal of a new kind. *Proc. K. ned. Akad. Wet.*, Ser. C **69**, 587–596.

MORSE, L. E. (1974). Computer programs for specimen identification, key construction and descriptive printing using taxonomic data matrices. *Publ. Michigan State Univ. Mus.*, Biol. Ser. **5**, 1–128.

OSBORNE, D. V. (1963). Some aspects of the theory of dichotomous keys. *New Phytol.* **62**, 144–160.

PANKHURST, R. J. (1970). A computer program for generating diagnostic keys. *Comp. J.* **13**, 145–151.

PANKHURST, R. J. (1972). A method for data capture. *Taxon* **21**, 549–558.

4 | Genkey: a Program for Constructing Diagnostic Keys

R. W. PAYNE

Rothamsted Experimental Station, Harpenden, Hertfordshire, England

Abstract: Genkey is a FORTRAN computer program for constructing diagnostic keys from a matrix giving the responses of each species to tests, together with certain ancillary information. The paper describes the methods used in Genkey and the options available in, for example, mode of construction, test selection, use of probabilities and forms of output.

Key Words and Phrases: keys, identification, probabilistic keys, test selection

INTRODUCTION

The aim in writing Genkey was to produce a program which is simple to use, with a wide range of options governing type of key, method of construction and form of output. This account omits details and concentrates on the main features, particularly the more novel ones. The program, like those written elsewhere, does not guarantee to produce the best key, for which algorithms are known only for very small problems or those with special features (e.g. binary tests* where each test gives a positive response for only one species). However, Genkey usually produces a good key which can be improved by the user with additional runs modifying the options and parameters.

MODE OF CONSTRUCTION

Tests (characters) are chosen sequentially without looking ahead to see if the current choice will turn out to be poor when subsequent choices are made. Tests may be selected either in branch-by-branch order—that is by choosing all the tests on each branch before going on to build the next; or in depth-by-depth

* In botanical or zoological applications, using a test is equivalent to determining the state of a character.

Systematics Association Special Volume No. 7, "Biological Identification with Computers", edited by R. J. Pankhurst, 1975, pp. 65–72. Academic Press, London and New York.

order—that is by choosing all the tests on each depth before considering the next depth down. Depth-by-depth construction is the more complicated and uses more computer time, but when previously used tests are preferred (see "selection of tests" below), this enables the tests towards the top of each branch to be chosen without bias from tests at the end of previous branches, and favours the subsequent use of those tests.

TYPES OF TESTS

Tests may have any number of levels (i.e. possible answers, or character states) and the program will also cope with situations where the result of a test on a species is either unknown or variable (in which case any of the possible answers may occur). These will be designated ?-responses.

Tests may be considered either singly or in combination (a multi-test). A multi-test is equivalent to a single test with many levels and selection is as for single tests, except that the best multi-test is chosen from all multi-tests up to the order specified by the user. However, the computer time required increases rapidly with the order of the multi-test, and tests of order more than 4 or 5, or for larger keys more than 2 or 3, are not recommended.

Costs of tests may be specified and used in the selection procedure (see below).

PROBABILITIES

Prior probabilities of the species (taxon weightings) may be specified. For each species this is the probability that an unseen sample is a member of the species concerned. Use of prior probabilities tends to reduce the number of steps required to identify the more likely, or common, species.

Probabilities may also be specified for the possible outcomes (levels) of the ?-responses. If one of these probabilities is zero then the species concerned will not appear on the branch produced by that answer. This also provides a tool for extending the meaning of ?-responses to species whose responses are variable over some but not all levels of a test.

SELECTION OF TESTS

When there are no ?-responses the test is selected which most nearly splits the species into equal groups (or when prior probabilities are being used the test is that which most nearly splits the species into groups of equal probability). When ?-responses are present, the method also aims to keep the number of ?-responses low without making the group sizes too disparate.

Two selection criteria are provided:

1. Circle criterion

For comparison of tests with up to n levels, minimize

$$(p_1 - 1/n)^2 + \ldots + (p_n - 1/n)^2 + r^2,$$

where p_i is the proportion (or probability) of species giving the ith answer to the test, and r is the proportion of ?-responses.

2. Shannon-information criterion

Minimize $(p_1 + r) \log (p_1 + r) + \ldots + (p_n + r) \log (p_n + r)$. When r is small and the p_i near to $1/n$ these will give very similar results, but as the p_i diverge from $1/n$, Shannon-information tends to choose tests with larger values of r than the circle criterion. Shannon-information also takes more computer time because it calculates logarithms. It would not be difficult to incorporate other criteria but none with obvious advantage seems to have been suggested.

To enable the user to express preference about which test should be used, a group of equivalent tests is formed, defined as those whose criterion value differs from that of the best test available by less than a specified amount. The test is then chosen from this group according to one of a set of rules of which the following are examples:

(a) Best of the tests that have occurred before (on another branch).
(b) Best of the tests with least cost.
(c) Best of the tests with the least number of ?-responses.

(b) and (c) may be further restricted to tests which have been used before.

The above are for use with single tests. When multiple tests are included, such rules as the following may be used:

(d) Best of the multi-tests with least (total) cost.
(e) Best of the multi-tests with least cost, the fewest of whose elements have not occurred before.
(f) Best of the multi-tests with least cost, of those with the fewest ?-responses.

Judicious use of these rules emphasizes different aspects of the resulting key. Thus (a) reduces the number of different tests used in the key, (b) reduces cost and allows zero cost tests to be selected early, (c) tends to reduce number of depths, etc. More details of test selection, with particular reference to the circle criterion, are given by Gower and Barnett (1971).

MULTI-SPECIES

The program caters for species which cannot be positively separated by the given tests (multi-species). Further separation may be made in two ways:

1. Use of ?-responses

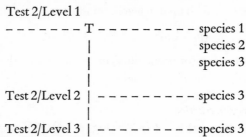

In the example above there is no test that will positively separate species 1, 2 and 3. However, when test 2 is applied, species 1 and 2 definitely give the first answer while species 3 has a ?-response. Hence, as can be seen above, should test 2 be applied giving either of answers 2 or 3, species 3 will be identified uniquely. The criterion used for selecting such tests is to maximize the number of (or probability of) species giving the ?-response (provided at least one species gives a definite response).

2. Probabilistic Key

In addition, if probabilities have been given for different levels of ?-responses, the user may request the program to construct a probabilistic key for the members of multi-species. The criterion used for test selection is to maximize

$$\sum_i \sum_j p_{ij} \log \frac{p_{ij}}{\sum_k p_{ik}}$$

where p_{ij} is the probability that species i gives answer j to the test concerned.

To ensure that branches do not continue unnecessarily, a cut off probability may be specified. A branch will terminate when one of the species remaining on the branch has a probability greater than that specified.

This method of separation may be provided in one of three ways: as part of the main key, as a separate key for each multi-species given after the main key, or as a list of tests, in order of usefulness, for the possible separation of each multi-species.

INPUT OF DATA

The data are read in sections each of which is identified by a directive, only the first three letters of which are significant. Numbers may be read either in fixed format, in which case a format statement must be given before each set of numbers, or in a free format which is simpler and more flexible.

As an example, if there are 40 tests each with two levels, in free format this may be specified by "levels 40(2)".

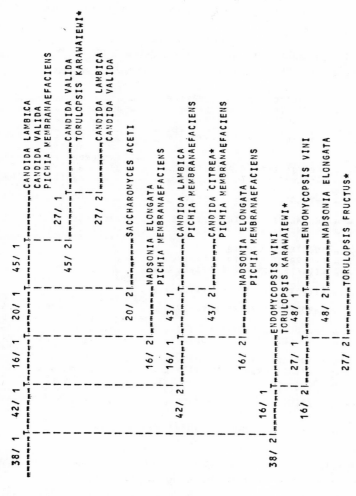

FIG. 1. Part of a key, in diagrammatic form, for 434 species of yeast. The notation 38/1 means test no. 38/level 1.

The results of the tests on each species are represented as non-zero positive integers and specified in species × test form using the directive "Results".

"Options", "Parameters", "Costs" and "Probabilities" are examples of other such directives.

Directives are also used to give instructions to the program e.g. Run, End.

OUTPUT

Examples of the possible forms of output are shown in Figs 1, 2 and 3, which are part of a key for 434 species of yeast, the data for which was provided by

```
1 ▪ ▪ ▪  D-GLUCITOL GROWTH NEGATIVE ▪ ▪ ▪ ▪ ▪ ▪ ▪ ▪ ▪ ▪ ▪ ▪2
  2 ▪ ▪ ▪  SUCCINATE GROWTH NEGATIVE▪ ▪ ▪ ▪ ▪ ▪ ▪ ▪ ▪ ▪ ▪3
    3 ▪ ▪ ▪  L-SORBOSE GROWTH NEGATIVE▪ ▪ ▪ ▪ ▪ ▪ ▪ ▪ ▪ ▪4
      4 ▪ ▪  TREHALOSE GROWTH NEGATIVE▪ ▪ ▪ ▪ ▪ ▪ ▪ ▪ ▪ ▪5
        ·5 ▪ ▪  ARBUTIN GROWTH NEGATIVE▪ ▪ ▪ ▪ ▪ ▪ ▪ CANDIDA LAMBICA
                                                     CANDIDA VALIDA
                                          PICHIA MEMBRANAEFACIENS
        5 ▪ ▪  ARBUTIN GROWTH POSITIVE▪ ▪ ▪ ▪ ▪ ▪ ▪ ▪ ▪ ▪ ▪ ▪6
          6 ▪ ▪  D-XYLOSE GROWTH NEGATIVE ▪ ▪ ▪ ▪ ▪ ▪CANDIDA VALIDA
                                               TORULOPSIS KARAWAIEWI*
          6 ▪ ▪  D-XYLOSE GROWTH POSITIVE ▪ ▪ ▪ ▪ ▪ ▪ CANDIDA LAMBICA
                                                     CANDIDA VALIDA
      4 ▪ ▪  TREHALOSE GROWTH POSITIVE▪ ▪ ▪ ▪ SACCHAROMYCES ACETI
    3 ▪ ▪ ▪  L-SORBOSE GROWTH POSITIVE▪ ▪ ▪ ▪ NADSONIA ELONGATA
                                          PICHIA MEMBRANAEFACIENS
  2 ▪ ▪ ▪  SUCCINATE GROWTH POSITIVE ▪ ▪ ▪ ▪ ▪ ▪ ▪ ▪ ▪ ▪7
    7 ▪ ▪ ▪  L-SORBOSE GROWTH NEGATIVE▪ ▪ ▪ ▪ ▪ ▪ ▪ ▪ ▪8
      8 ▪ ▪  CITRATE GROWTH NEGATIVE▪ ▪ ▪ ▪ ▪ ▪ CANDIDA LAMBICA
                                          PICHIA MEMBRANAEFACIENS
      8 ▪ ▪  CITRATE GROWTH POSITIVE▪ ▪ ▪ ▪ ▪ ▪ CANDIDA CITREA*
                                          ·PICHIA MEMBRANAEFACIENS
    7 ▪ ▪ ▪  L-SORBOSE GROWTH POSITIVE▪ ▪ ▪ ▪ ▪ NADSONIA ELONGATA
                                          ·PICHIA MEMBRANAEFACIENS
1 ▪ ▪ ▪ ▪  D-GLUCITOL GROWTH POSITIVE ▪ ▪ ▪ ▪ ▪ ▪ ▪ ▪ ▪ ▪9
  9 ▪ ▪ ▪  L-SORBOSE GROWTH NEGATIVE▪ ▪ ▪ ▪ ▪ ENDOMYCOPSIS VINI
                                               TORULOPSIS KARAWAIEWI*
  9 ▪ ▪ ▪  L-SORBOSE GROWTH POSITIVE▪ ▪ ▪ ▪ ▪ ▪ ▪ ▪ ▪ ▪ 10
    10 ▪ ▪ ▪  D-XYLOSE GROWTH NEGATIVE ▪ ▪ ▪ ▪ ▪ ▪ ▪ ▪ ▪ 11
      11 ▪ ▪  NO VITAMINS GROWTH NEGATIVE▪ ▪ ▪ ▪ ▪ ENDOMYCOPSIS VINI
      11 ▪ ▪  NO VITAMINS GROWTH POSITIVE▪ ▪ ▪ ▪ NADSONIA ELONGATA
    10 ▪ ▪ ▪  D-XYLOSE GROWTH POSITIVE ▪ ▪ ▪ ▪ TORULOPSIS FRUCTUS*
```

FIG. 2. Conventional form of key shown in Fig. 1.

Dr J. A. Barnett, University of East Anglia, Norwich. A full key for these yeasts is given by Barnett and Pankhurst (1974). Figure 2 may also be printed without indentations if required. Figure 3 is a new representation fully described by Payne et al. (1974). At the end of each branch in Fig. 1, the cost of using the tests on that branch may be printed, and also the probabilities for members of multi-species. The program will also print the average (or expected) cost of identifying an unknown sample. Inside the computer the key is held in a compact coded form. This may be printed and/or output to a line file whence

	TEST	NEGATIVE	POSITIVE
1	D-GLUCITOL GROWTH	2	9
2	SUCCINATE GROWTH	3	7
3	L-SORBOSE GROWTH	4	NADSONIA ELONGATA
4	TREHALOSE GROWTH	5	PICHIA MEMBRANAEFACIENS
5	ARBUTIN GROWTH	CANDIDA LAMBICA CANDIDA VALIDA PICHIA MEMBRANAEFACIENS	SACCHAROMYCES ACETI
6	D-XYLOSE GROWTH	CANDIDA VALIDA	CANDIDA LAMBICA CANDIDA VALIDA NADSONIA ELONGATA PICHIA MEMBRANAEFACIENS
7	L-SORBOSE GROWTH	TORULOPSIS KARAWAIEWI*	8
8	CITRATE GROWTH	CANDIDA LAMBICA PICHIA MEMBRANAEFACIENS	CANDIDA CITREA* PICHIA MEMBRANAEFACIENS
9	L-SORBOSE GROWTH	ENDOMYCOPSIS VINI TORULOPSIS KARAWAIEWI*	10
10	D-XYLOSE GROWTH	11	TORULOPSIS FRUCTUS*
11	NO VITAMINS GROWTH	ENDOMYCOPSIS VINI	NADSONIA ELONGATA

FIG. 3. Alternative form of key shown in Fig. 1. Each test name appears once followed by all its possible levels.

it may be read by the program at a later date and, in conjunction with the data used in constructing the key, additional copies or new forms may be printed. The coded key is also useful for providing ancillary tables of tests used, numbers of multi-species, etc.

The program is written in FORTRAN IV and mounted on the Rothamsted ICL 4.70 computer, but transfer to other computers would not present insuperable difficulties. Further information, including a user's guide, may be obtained from the author.

REFERENCES

GOWER, J. C. and BARNETT, J. A. (1971). Selecting tests in diagnostic keys with unknown responses. *Nature, Lond.* **232**, 491–493.
BARNETT, J. A. and PANKHURST, R. J. (1974). "A New Key to Yeasts". North-Holland, Amsterdam.
PAYNE, R. W., WALTON, E. and BARNETT. J. A. (1974). A new way of representing diagnostic keys. *J. gen. Microbiol.* **83**, 413–414.

5 | A Computer Program to Construct Polyclaves

R. J. PANKHURST* and R. R. AITCHISON

Department of Botany, University of Cambridge, England

Abstract: A diagnostic key on punched cards (a polyclave) can be manufactured by computer. The construction and use of polyclaves is described, together with some applications. A polyclave is used entirely in the hand, and when prepared by computer, is a cheap and versatile identification method with great potential.

Key Words and Phrases: identification, polyclave, taxonomic data matrix, Rubus, yeasts

INTRODUCTION

The term "polyclave" was coined by Duke (1969). We take it to mean a form of diagnostic key which is expressed as cards (or some other material) which are laid on top of one another in order to eliminate taxa which disagree with the specimen to be identified. This is a wider sense than intended by Duke. Each card corresponds to a character-value combination, as in a simple lead of a printed key. A polyclave can also be called a "multiple-entry" key. Morse (1971) used the term polyclave in a wider sense, to include, for instance, on-line identification programs, whose logic is similar, but whose implementation is different. Morse gives references to various forms of polyclave (*sensu stricto*) such as "peek-a-boo" cards and window keys. The familiar edge-punched library index cards are a form of polyclave. There exist many punched card polyclaves which are expressed on specially manufactured cards, or which were punched by hand. As an example of this we may quote the key to world angiosperm families by Hansen and Rahn (1969). The polyclave of Duke (1969) consists of sheets of transparent plastic with small scale printing in black which are made to overlie a white card printed in red, but most polyclaves have cards with punched holes, around the margin or all across the card.

---

* Present address: *British Museum (Natural History), London.*

Systematics Association Special Volume No. 7, "Biological Identification with Computers", edited by R. J. Pankhurst, 1975, pp. 73–78. Academic Press, London and New York.

DESCRIPTION AND USE OF THE POLYCLAVE

The following remarks apply specifically to the polyclaves produced by our program. The standard card has 80 columns and 12 rows, so that 960 positions can be punched with rectangular holes.

A polyclave consists of a pack of computer cards each one of which represents a character value, e.g. petals white. The character value is written at the right-hand end of the card. Cards are also numbered at the right in a more or less alphabetical order by character and should be replaced after use so that they can be found again later. At the left-hand end of each card is a series of holes. Each hole represents a species which shows that character state, e.g. in a flowering plant key a hole in column 14, row 6 in a card inscribed "petals pink" indicates that species no. 146 has pink petals (N.B. it could show other petal colours too).

To use the polyclave, cards are chosen from the pack, one at a time, in any order, to represent the character values shown by the specimen to be identified. Only one card can be used for each character, e.g. if your specimen really has both pink and white petals that character must be ignored. As each card is selected the chosen cards are overlaid and held up to the light. Places where holes still show at the left-hand side are those species which have that selection of characters. This process is continued until one hole remains. The number of this species is then looked up on the list of species names and numbers provided with the polyclave and the specimen checked against the literature description and museum specimens, if available. (Numbering starts at 11 because there is no column 0 on the cards.) If no holes are left then the cards chosen should be reassessed and those which the user is not absolutely sure about should be discarded and, if possible, new cards added. If there is a large number of holes left the best way to reduce this is to select an uncommon character value shown by the specimen. An "uncommon" character state is one which has few holes at the left-hand end of the card.

Obviously the more that "uncommon" characters are used the quicker will be the reduction in the number of species remaining. However, the specimen may show very few or none of these "uncommon" characters in which case care must be taken in the selection of cards.

Characters are selected entirely in the order that the user wishes so that he can use, for example, all flower or all leaf characters. However, some characters are not good ones when used in the polyclave because they are either unreliable or very variable. "Middle" states of quantitative characters seem particularly suspect, e.g. "prickles fairly many" is less sure than "prickles few" or "prickles many", because of borderline cases. If the specimen shows a doubtful or variable character it is best to avoid using this unless it is really necessary. Other "bad" characters will be discovered with experience.

Geographical distribution cards may be available, and where appropriate, can convert a general key into a specialized one, e.g. card for records in Cambridge-shire converts an all Britain key to a local key. But the user must beware of using these—he may have a hitherto unrecorded species.

Although polyclaves may be a novelty, the user must realize that they are still subject to some of the disadvantages of conventional keys, e.g. one may find that a correctly identified specimen still disagrees in several characters from the key, especially for critical groups.

As well as its use as a key, the polyclave can be used to find a unique set of characters for a given species by selecting cards which have holes in the appropriate column and row number. Cards can then be taken away one at a time until a minimum number are left showing one hole at the appropriate place.

This set of cards then shows a unique set of characters to identify the specimen. However, great care must be taken *not* to include character state cards for those characters which are variable or unknown (see above), e.g. if the species has holes in cards "petals pink" and "petals white" then the character petal colour should not be used for this purpose. If this type of character is used then the set of characters is not unique. This system is quite a help in the field when looking for a particular species among several of its close relations.

CONSTRUCTION OF THE POLYCLAVE

The cards are punched by a FORTRAN program, which takes as input a taxonomic data matrix and a list of the character names and their associated value names. The taxonomic data matrix has the same form as that used by the matching program (Pankhurst, this volume, Chapter 6). Descriptions of taxa are represented by integer numbers. Variable characters are included by coding the states by powers of two and adding them together. For example, the colours white, pinkish, pink and dark pink are represented by 1, 2, 4 and 8 respectively, rather than by 1, 2, 3 and 4. This is because the resulting sum uniquely represents the values, e.g. value $6 = 2 + 4$ only, and not any other combination of powers of 2. This device enables a variable character to be described in the same amount of computer storage as a constant one.

The program is as near standard FORTRAN as possible, but the details of punching cards with an arbitrary pattern of holes ("binary" cards) are bound to be machine-dependent. In this case, working with an IBM 370/165 and the standard operating system, it is firstly a matter of having the correct "job control language" (JCL) cards. With this machine, one half-word of store is used to punch one column on a card, although with other machines the arrangement might be otherwise. The bit (binary digit) pattern in this word determines which holes are punched, one bit per hole. It is therefore necessary to discover what integer number corresponds to the store contents for producing a particular hole, and to add these numbers together for each hole required in the column.

When the value of a character is variable, then it is clear that a hole must be punched in the appropriate place on each of several cards representing the different values. Likewise, if a character value is missing, it will generally be reasonable to assume that any of the possible values might occur, and to punch a hole in all of the appropriate cards. Although it would be misleading to attempt to read back a taxon description from cards punched this way, the actual occurrence of one of these states on a specimen could eliminate a taxon wrongfully unless all the holes are punched. A character which is conditional on another character may be missing because it is impossible to observe, but, even here, punching all holes will not affect identification. An exception concerns data relating to distribution information, where a missing value means that the taxon has not been recorded in any of the relevant area, so there must be no holes punched. Whether a missing value is to be punched or not is determined by data in the list of characters, and it is usual to assume that missing characters will be punched. In order to decompose a combined value which expresses a variable character into its component values, it is convenient to carry out an "and" (logical product) operation, expressed by a suitable special subroutine. This could also be done by dividing successively by 2 and looking for remainders, but this would be wasteful of computer time. This is the only other point where the program is machine dependent.

When the polyclave cards are in use, cards have to be found and replaced in the deck. In order to find them, we have usually arranged the characters in alphabetical order. In order to put them back easily, the cards are numbered at the right hand end. The sequence number is the combination of a character number and a value number: e.g. the third card for character 22 is labelled 223. In this way it is easy to see which cards belong together for one character. It has frequently been the case that the order of cards in the polyclave desired by those who use them does not correspond to the order of characters in the data matrix, so the program allows for this.

The manufacture of numerous copies of a polyclave can be carried out either by storing the card images on magnetic disk or tape and running a program to copy them, or by putting the cards through a separate copying machine (reproducer). In the latter case, different card colours can be used within one polyclave. This has been found useful in order to draw the user's attention to particular cards, e.g. those for characters recommended as most reliable. The same effect is also achieved by colouring the edges of the cards with a felt pen. The cards which are manufactured as above emerge without the characters printed on them, and they are "interpreted" by a card punch machine which reads the cards and prints their contents on the top margin. Part of each card is unprintable since it contains non-character information, so the interpreting machine has to be set up (programmed) to skip over part of the card. A printed list of the taxa by number, which has to go with the card deck, is easily manufactured on a computer printer.

Several other ideas were tried out in order to improve the polyclave. It was thought that if a specimen happened to show an uncommon value of a character, then it would be useful to know what other characters showed further un-common values in combination with the first. This would lead to rapid recognition of distinctive taxa. A program was written to explore the data matrix for uncommon pairs of character values, i.e. those pairs which occurred with a frequency of 10% or less. These pairs were cross-referenced on the cards, so that if, say, character 23 value 2 occurred rarely in combination with character 43 state 1, card 232 would have a comment "see 43" and card 431 would be marked "see 23". Trials suggested, however, that combinations which were rare enough to be useful occurred too rarely for it to be worthwhile searching for them.

Another promising idea, suggested by A. O. Chater, is an "inverse" polyclave. This is an additional polyclave, with a card for each taxon. When two cards for different taxa are held together, they will answer the question, "What are the differences between these two taxa?". This would be useful because it sometimes happens that a small number of taxa remain to be distinguished, and much time can be spent in searching for diagnostic characters.

OTHER WORK

The computer-produced polyclave described here is not quite the first to appear, since Morse (1974) has recently constructed one, and so have Weber and Nelson (1973). An ingenious idea for improving a polyclave has been tried out by R. W. Rayner (unpublished), who used tinted transparent plastic sheet, punched by hand. In this way, a few mistakes in character interpretation will not necessarily prevent a correct result being reached, because the taxa which agree most will show up as the brightest areas, even if they are not clear right through. Rayner has reported difficulty, both with obtaining material of uniform tint, and with choosing the correct shade for packs of different thickness. Whether 80-column cards are available in such material is not known.

APPLICATIONS

A number of polyclaves have been constructed. The first and largest was for 400 microspecies and 118 characters of *Rubus fruticosus* in Great Britain (Pankhurst, 1973). This is an especially troublesome group, and the polyclave cannot be evaluated by comparison with a printed key, since the latter does not exist. A polyclave for all known species of yeast (434) with 60 characters was prepared with the help of J. A. Barnett (Barnett and Pankhurst, 1974). Students helped to prepare a polyclave to certain species of *Silene* as part of a course on taxonomy, and an ecologist who is working on tropical montane forest in Jamaica will use a polyclave to help distinguish about 80 tree species. A polyclave to fossil pollen of quaternary age is being evaluated, and one to diseases of the oesophagus has been shown to perform as well as a medical expert (Edwards and Pankhurst, 1974).

CONCLUSIONS

A polyclave is in the same general class of identification method as a diagnostic key, but is more flexible since any characters can be used in any order. Now that polyclaves can be produced on computer cards, it is easy and cheap to manufacture them in quantity, and they are also easy to correct and re-construct. A large number of taxa (perhaps up to 600) can be accommodated. Because a computer is only needed for manufacture, polyclaves can be applied in under-developed areas, particularly in the tropics, where taxonomic work is the most urgent.

REFERENCES

BARNETT, J. A. and PANKHURST, R. J. (1974). "A New Key to the Yeasts". North-Holland, Amsterdam.
DUKE, J. A. (1969). On tropical tree seedlings. I. Seeds, seedlings, systems and systematics. *Ann. Mo. bot. Gdn.* **56**, 125–161.
EDWARDS, D. A. W. and PANKHURST, R. J. (1974). Development of a diagnostic key for dysphagia. Proc. Medical Data Processing Symposium, Toulouse, March 1974.
HANSEN, B. and RAHN, K. (1969). Determination of angiosperm families by means of a punched-card system. *Dansk bot. Ark.* **26**(1), 46 pp.
MORSE, L. E. (1971). Specimen identification and key construction with time-sharing computers. Taxon **20**, 269–282.
MORSE, L. E. (1974). Computer programs for specimen identification, key construction and description printing using taxonomic data matrices. Publ. Mus. Michigan State Univ., Biol. Ser. **5**, 1–128.
PANKHURST, R. J. (1973). Polyclave for *Rubus fruticosus* agg. in Great Britain, Internal Report, Department of Botany, University of Cambridge.
WEBER, W. A. and NELSON, P. (1973). Random access key to the Genera of Colorado Mosses. Review by D. H. Vitt, *Bryologist* **76**(1), 222–223.

6 | Identification by Matching

R. J. PANKHURST

Department of Botany, University of Cambridge, England*

Abstract: A method is described for identification by matching a specimen to all possible taxa by computer. The example used is the apomictic microspecies of *Rubus* in Great Britain. Various different ways of calculating similarity were explored, and efforts were made to reduce the number of characters which need to be examined. Details are given of the performance of the program, which is able to indicate the correct identification with more than 80% of the specimens tested.

Key Words and Phrases: identification, matching, special characters, Rubus, similarity coefficient, character weighting, conditional characters, character selection, separation coefficient

INTRODUCTION

The method of identification by matching is very simple in principle. A specimen is compared with every possible taxon with which it could be identified, and a measure of likeness computed in each case. A proportion of those taxa with the highest score is used as a short list, from which the final determination is made by some other means. In this instance, a similarity coefficient is used to estimate likeness, and the top 10% of the scores appear in the output.

This technique is one of those basic to pattern recognition (Nagy, 1968), but applications of it in biology are few (Walker *et al.*, 1968; Hall, 1969; Hansen and Cushing, 1973; Campbell, 1973; Ross, this volume). In this paper, an application to the genus *Rubus* in the British Isles is discussed in full. *Rubus* taxa are largely apomictic and numerous "microspecies" have been described. They were chosen because:

(a) the identification problem is unusually severe;
(b) they are shrubs, abundant and widespread, so that access is easy;
(c) a good herbarium was available.

* Present address: *British Museum (Natural History), London.*

Systematics Association Special Volume No. 7, "Biological Identification with Computers", edited by R. J. Pankhurst, 1975, pp. 79–91. Academic Press, London and New York.

DESCRIPTION OF PROGRAM OUTPUT

An example is shown in Fig. 1. The title line is a comment with a tentative name (R. PLICATUS) and details of the collection. Figures are given for the comparison between the specimen and 40 species, representing 10% out of a total 400, and a sequence number is printed in the first column. The second column gives the estimated similarity expressed as a percentage, and the third column the number of characters known in both the specimen and in the stored description of the species, on which the similarity is based. The taxon names are given at the right, after a series of special symbols (asterisk and plus signs). If any specimen and taxon have too few characters between them for comparison, the taxon name is printed in a warning message. The limit is set at 25 characters in this application.

The asterisk marks an attempt to assign the specimen to the correct group of taxa (or taxon of higher level), rather than the correct individual taxon. This is intended as a means to help make a choice from the short list of identifications already presented. The taxa are divided into a convenient number of groups (about 10 in this case) according to some classification or clustering, and an asterisk is printed for each member of that group which has the highest average similarity with the specimen.

At the head of the output is a list of special characters (e.g. length of petals) which are chosen by the user as an option. The notion is that certain character states of the specimen are often striking to the observer, such that he would have difficulty in accepting any determination which did not agree in most or all of these respects. A plus sign is printed against the name for each of these special characters whenever the taxon in question agrees. For example, the first taxon, SCISSUS, agrees with all three special characters.

A further option allows particular tentative identifications to be explored directly. In the example, *R. plicatus* was requested and the program prints its sequence number and percentage similarity under the heading "special taxa". This is for the case when there already exists some reason to suppose that the specimen might belong to a certain taxon, and one wishes to be sure of having its similarity printed.

With this information, one can investigate the identity of the specimen further. Taxa which are asterisked and which agree in all the special characters are an obvious starting point. In the example, there are three of these, namely: 1. SCISSUS, 5. OPACUS and 10. PLICATUS. On other evidence, PLICATUS is taken to be the correct answer. Otherwise, one might investigate those asterisked but with fewer pluses e.g. 18. DIVARICATUS, or those not asterisked with three pluses e.g. 14. DURESCENS, and so forth. Taxa which

```
/R.PLICATUS,SDATA 56,OPEN,S OF CAR PARK,DANBURY CMN,18.7.1973/
SPECIAL CHARACTERS ARE -
     STEM/     HABIT
     PETALS/LENGTH
     PANICLE LENGTH/


SEQ     SIM.  COUNT                 SPECIES

  1     65.2   102    *+++          SCISSUS/
  2     65.2   103    *+ +          INTEGRIBASIS/
  3     65.1    93      +           DOBUNIENSIS/
  4     64.9   104    *+ +          ARRHENIIFORMIS/
  5     64.4   103    *+++          OPACUS/
  6     63.6    96                  BROENSIS/
  7     63.4    96      +           CHLOOCLADUS/
  8     63.1   103                  LATIFOLIUS/
  9     63.0    94      +           OOLITICUS/
 10     62.8   102    *+++          PLICATUS/
 11     62.7    94      +           AMPHICHLORUS/
 12     61.5   104    *+ +          AMMOBIUS/
 13     61.5   100      +           HIRTIFOLIUS/
 14     61.5   103     +++          DURESCENS/
 15     61.4   103     +++          NEOMALACUS/
 16     61.2   103    + +           HOLERYTHROS/
 17     61.1   103    + +           GRATUS/
 18     61.1   105    *+ +          DIVARICATUS/
 19     60.8   104                  LEUCANDRUS/
 20     60.8    93      +           MEGAPHYLLUS/
 21     60.7   104    *+ +          SULCATUS/
 22     60.7   103      +           POLIODES/
 23     60.6    99    *             MERCICUS/
 24     60.3   104      +           AMPLIFICATUS/
 25     60.2   103                  GLANDULIGER/
 26     60.1    94     ++           DIVERSIARMATUS/
 27     60.1    97      +           HESPERIUS/
 28     60.1    99                  OBSCURIFORMIS/
 29     60.0   103                  SILURUM/
 30     59.8   105     ++           CARPINIFOLIUS/
 31     59.8   105     +            NITIDOIDES/
 32     59.7    94                  HORRIDISEPALUS/
 33     59.7    91      +           UNCINATIFORMIS/
 34     59.6    96                  LIBERTIANUS/
 35     59.0   102     +++          CHLOOPHYLLUS/
 36     59.0   103     +            PERMUNDUS/
 37     58.8    98      +           ATROCAULIS/
 38     58.8   103                  CONFERTIFLORUS/
 39     58.7   103                  CAMBRENSIS/
 40     58.7   102                  PULLIFOLIUS/


SPECIAL TAXA COMPARED
 10     62.8                        PLICATUS/
```

Fig. 1. Output from matching program.

have a lower count of comparable characters than the rest, e.g. UNCINATI-FORMIS, with 91, should be treated with caution. The similarity may be more in error (too high or too low) since fewer characters were used. Notice that the similarity for the "correct" answer was 62%, as opposed to 65% for the "best" answer. Owing to errors of observation both in the specimen and in the reference data, and natural variation, the highest similarity is not necessarily that of the "right" answer. The higher the similarity, the more likely is the determination, but the similarity itself should not be regarded as a probability. With this particular application, highest similarities of about 70% are usual, and over 80% is rare. These figures may seem low, but are believed to be explained largely by natural variation.

DESCRIPTIVE DATA

The data on which the identifications are based were derived, in the first place, entirely from a monograph (Watson, 1958). Initial experiments with these data were disappointing, giving a correct result for only about one specimen in two even with distinctive and easily recognized taxa (see Table III). Changes in the manner of calculation of the similarity, as described below, brought no improvement. It was found that a marked improvement in the results was obtained by filling in missing data, not supplied in the monograph, from herbarium specimens. The manner in which the data were completed is described in this volume in another paper (Pankhurst, Chapter 14). It is important to note that data from different sources were all added together to form combined descriptions, in such a way that the distinction between variation within and between specimens was lost. This was done to avoid unduly increasing the number of taxa which have to be considered by the computer. There is evidence, explained below, that this approximation does not materially affect the results of the matching program.

PROGRAM DATA

The data currently consist of a matrix of 118 characters for 400 species. These data were obtained initially from Watson's monograph. Tests of the program on these data showed poor performance; only about one in two of the "correct" names for the specimens tested appeared in the output lists. Investigation of a variety of different ways of calculating the similarity coefficient (described below) revealed no improvement. Only when the data were improved by adding missing information did any improvement show. This was done by selecting a specimen of *Rubus ulmifolius* (a common and distinctive species) and adding extra data which had been missing for this species plus those that appeared in the output list before it (20 of the total of 400). *R. ulmifolius* then moved from 20th in the list to first when the program was run again. It was then decided to complement the whole of the existing data by describing specimens from a herbarium collection. This process itself gave rise to a considerable number of problems, whose solution is described in the paper on the quality of taxonomic descriptions (Pankhurst, this volume, Chapter 14). In this procedure, information about one species from various sources was combined together to give one description per species. This means that information about variation between different specimens of the same species is lost. This was done because the time for the computer program to run is proportional to the number of taxa considered, so it is best not to let this number grow too large.

When a character of a taxon is variable, a special method is used in order to store the variable states in just one computer word, as in the case where the character is constant. Consider the character "stem colour", which has states "yellow", "green", "red" and "brown". These states could be numbered from 1 to 4, from left to right. A stem which is both green and red shows states 2 and 3. This would need two computer words to store the two numbers, plus another one to record the fact that there are two states present. If the states are numbered in powers of two, e.g. 1, 2, 4 and 8, "red" and "green" becomes 2 plus 4, i.e. 6. There is only one way to reduce 6 to a sum of any number of integers belonging to the series 1, 2, 4, 8 etc., so the number 6 is stored in one computer word, and still represents both states. When this is done throughout, space is saved and simplicity is achieved.

THE SIMILARITY COEFFICIENT

A similarity coefficient is used, rather than the converse (a dissimilarity coefficient), because it is thought more natural in identification. The different forms of the coefficient described here were tried out with both the initial and completed versions of the data, but the effect of calculating the coefficient in different ways was found to be much the same for either of the two matrices. The coefficient is a simple similarity coefficient (Sneath and Sokal, 1973), defined as:

$$\frac{\text{number of characters which agree}}{\text{total number of characters}}$$

This coefficient includes negative matches i.e. if a character state is of the form "absent" in both the objects being compared, then this is counted towards the agreement. This is thought appropriate here because the taxa involved are rather homogeneous in their descriptions, and problems of homology are negligible, i.e. there is no difficulty in deciding which organs correspond between different taxa.

The details of the calculation follow. Since the number of states per character was rather variable (from 2 to 7), the characters were weighted according to the logarithm to the base 2 of the number of states. This effectively reduces all characters to binary characters when they are combined in the similarity coefficient. For example, a 4-state character is weighted by $\log_2 4 = 2$, because it is equivalent to two 2-state characters. Notice that this is not a matter of unequal *a priori* weighting, but a matter of reducing existing known bias and removing effectively unequal weighting caused by unequal numbers of states. It was thought best to continue to use the characters with unequal numbers of states because of convenience for description purposes.

The coefficient was adapted to allow for variable characters when comparing a specimen with a known taxon. This is best illustrated by an example. Consider "petal colour", with states "white" (1), "pale pink" (2), "pink" (3), and "dark pink" (4). Various possible states are shown in Table I. This may be

TABLE I. Similarity coefficient for variable character

Specimen value(s)	Taxon value(s)	Agreement
0	1	0
2	2	1
1, 2	1, 2	1
1	2	0
1, 2	3, 4	0
1, 2	2, 3	1/3
1, 2	1, 2, 3	2/3

summarized as follows. If either or both values are missing, agreement is nil. If both values are identical, or if both ranges of values are identical, agreement is one. If values are different or if there are no values in common agreement is nil. If each shows a range of states, but these partly overlap, then the agreement is

$$\frac{\text{no. of states in common}}{\text{total no. of different states shown by both}}$$

In computer terms, using the method of storage described above, this is:

$$\frac{\text{no. of bits in the "and" of the two characters}}{\text{no. of bits in the "inclusive or"}}$$

Since the number of bits in a computer word is rather tedious to compute, this was stored in a table and retrieved by indexing an array as required. If the agreement between two taxa j and k for character i is called A_{ijk}, defined as above, with weighting w_i for character i, then the similarity between j and k is:

$$\frac{\sum_i w_i A_{ijk}}{\sum_i w_i}$$

Other variations on this coefficient were tried out, and the results are given in Table II. A set of 50 specimens was used for this, where these specimens had been named in the conventional way by experts. In each case, the number of

TABLE II. Performance of program and comparison of different coefficients on 50 species

	Simple		Inclusion		Near miss		Probability	
	no.	%	no.	%	no.	%	no.	%
first in list	7	14	9	18	5	10	6	12
in first 10	31	62	28	56	27	54	26	52
in first 20	37	74	34	68	35	70	32	64
in first 40	45	90	41	82	45	90	41	82
other	5	10	9	18	5	10	9	18

"correct" identifications by program, if appearance in the first 40 names in the results is taken to mean correctness, runs at between 80 and 90%, and since there is little to choose between them, the simple coefficient was used henceforth, as it is the easiest to calculate.

The coefficient with "inclusion" means that, whenever any character state from the specimen corresponded with any state for that character in the data, then full agreement was assumed. This may be regarded as assuming that all character variation occurs between specimens. The "near miss" version is an attempt to allow for errors of interpretation in observation. If specimen is "pilose" when a reference taxon is "pubescent", then agreement of half the full amount is allowed. This is only done for characters with three or more states and where these states can be put in ascending or descending order. Reduced agreement was only allowed for the immediately neighbouring state. The coefficient with "probability" is an attempt to make use of the fact that some character states are rarer than others. If a specimen shows a rare state, then this ought to be given more importance in the identification. Each character state was weighted by the negative logarithm (base 2) of its frequency, e.g. a state which only occurs in one taxon out of eight has a weight $-\log_2(\frac{1}{8})=3$. Since $-\log(0)$ is infinite, no state was given a weight of more than 10. Variable characters were allowed for by not counting them when finding the frequency of individual states. This approach may seem a promising one, but the results are no better than any other version. This is perhaps due to errors being magnified.

Another variation which was tried out was an effort to allow for conditional characters. For example, if a specimen has glandular hairs on the stem, then taxa in the data with "stem glands absent" will score nothing on this one character. However, other taxa with "glands present" can then also agree in respect of gland length, relative length, colour, and so forth. It might therefore be more accurate to weight "stem glands absent" by the number of other characters which depend on this. Eleven out of 118 characters showed this effect in *Rubus*, but the program results were much the same when the extra weights were included.

REDUCTION OF NUMBER OF CHARACTERS

Anything like a complete description of a specimen requires a large number of characters to be recorded. If one could obtain an identification on a reduced set of characters, then this could save a lot of work. Two ways in which this was attempted are described below.

(*a*) *By computation.* The method is approximate, because of vast amounts of computer time which would be required to solve the problem otherwise. To begin with, the aim is to find a set of characters which will enable every taxon to be distinguished from every other by at least one character. In order to decide on which characters to try and what order to try them in, a separation coefficient was calculated for each character, defined as

$$\frac{\text{no. of pairs of taxa actually separated by this character}}{\text{total possible no. of pairs}}$$

For N taxa, the total no. of pairs is $^NC_2$ or $N(N-1)/2$. The character with the highest separation coefficient is taken first, and then in order, until the number of unseparated taxa drops to zero. Notice that two taxa may be separable even if a character is variable, provided that the variation does not overlap. The order of characters obtained from the separation coefficient is a measure of their "goodness" for identification, but there is no allowance for correlation between characters. Also, multi-state characters tend to score higher than binary characters, because they really contain more information about the taxa.

Once a set of characters is known which makes every taxon distinct, it is worthwhile to continue to find more characters so that those pairs of taxa which are most similar are further separated. This is particularly true when no one character is completely reliable, and "spare" characters are needed for confirmation. A separate program was used for this purpose. At the point where every taxon pair is distinct, there will be at least some pairs which are distinguished only by one character. The program finds these pairs and finds those characters, so far unused, which would enable further distinctions to be made. The output is examined by hand in order to find a set of additional characters which is reasonably small. There are still a large number of ways to do this, and since it is a difficult computational problem, this selection is best done manually. At this stage a set of characters is known which will distinguish every taxon pair in at least two ways. This process is repeated until the point where this minimum difference can be increased no further. With the current *Rubus* data, 27 characters were sufficient to distinguish every taxon pair. In order to have at least 2, 3 or 4 differences 32, 42 and 55 characters, respectively, were needed. The minimum of 4 differences was as great as the data would permit. With 42 and 55 characters, 3 out of 42 and 4 out of 55 give about the same proportion of about 7% of the characters. This suggests that some taxonomic data might permit selection of characters in such a way as to provide deliberately a sufficient safety margin against errors in specimen data when identifying by matching. The 7% figure obtained here is probably not enough.

(*b*) *Subjectively*. After 3 years' experience of trying to identify *Rubus* specimens in the field, I felt able to make a character selection on the basis of ease of observation and confidence of accuracy. A set of 32 characters was selected in this way.

Table III shows the comparison of the results for the full set of characters and two reduced sets, one obtained by computing and the other subjectively. The specimens chosen for this comparison were chosen because they had been correctly identified by the matching program with the full character set. The set of 27 chosen by method (*a*) performs noticeably less well than the set of 32 chosen via method (*b*). This is probably because ease of observation of a character contributes much more importantly to correct results by avoiding errors than does the discriminating power alone. The difference between the actual numbers of characters (27 and 32) is probably not important. The 32 character set, when

88 R. J. Pankhurst

TABLE III. Comparison of performance using different character sets on 40 selected specimens

Position of correct name

		1st in list	In first 10	In first 20	In first 40	Elsewhere
118	no.	9	24	37	40	0
characters	%	22	60	92	100	0
32	no.	7	23	32	34	6
characters	%	17	57	80	85	15
27	no.	4	12	20	29	11
characters	%	10	30	50	72	28
Watson	no.	1	8	12	21	19
data	%	2	20	30	52	48

tested on the programs for method (a), turned out to be insufficient to distinguish all taxon pairs, but is better for matching in spite of this. An estimated 15% drop in performance between 118 and 32 characters is small in comparison with the reduction in the number of characters (nearly 75%). This shows clearly that character set reduction is well worthwhile. When using a matching program, it may well be sensible to score only the reduced set of characters for the first run, and then score the remainder later if necessary. However, since nearly half the set of 32 characters can only be observed on fresh material in the case of *Rubus*, this particular set would be unhelpful for preserved specimens of *Rubus*.

IDENTIFICATION BY GROUP

The program attempts to assign a specimen to the right group of taxa, as well as to the correct individual taxon, by printing an asterisk against all taxa in that group which are the most similar on average (see Fig. 1). The grouping was assigned initially by using the classification by "Sections" given by Watson. Out of a test set of 40 verified specimens, only 20 of the correct identifications were asterisked in the output. In an attempt to improve this, automatic clustering techniques were applied to the data matrix. Similarity coefficients were calculated using the simple coefficient described above. A single link clustering program due to Sibson (1973) was tried first. Groups deduced from this were tested as before, giving only six correctly asterisked results. An average link clustering program due to H. J. B. Birks (unpublished) was also used to obtain groups, and this gave 19 out of 40 correct. It seems that single link clustering is not useful for this purpose, and that the Watson classification is about as good as the average link clustering. It is probable that, from examination of dendrograms, apart from about 50 taxa which fall into two natural groups, the remainder represent a continuous spectrum of variation rather than a set of taxa which can conveniently be classified in a hierarchy. If this is so, it could explain why the "asterisk" feature is not very successful when applied to *Rubus*.

PERFORMANCE

The figures in Tables II and III, already referred to, show that the performance of the matching program on a notoriously difficult genus is encouraging. The program gives the correct result within the first 10% of possible taxa in 80–90% of cases. This is not as useful as it might be, however, when one considers that 10% of 400 still leaves a list of 40 names to search through, although this is further reduced by the use of special characters and the "asterisk" feature. It is also worth remarking that in the genus in question, a great many temporary forms of recent hybrid origin occur in nature, so that a substantial proportion of wild plants are unnameable, in the sense that it is not (and perhaps never will be) worthwhile to classify them. The computer time for each specimen is approximately one second on an IBM 370/165, in a FORTRAN program requiring 180K bytes of store, which is satisfactorily rapid.

Some experiments were also carried out on specimens which had been determined by experts as hybrids. In each of six cases, the program put one of the presumed parents in the first 10%, while the other was placed much lower. This suggests that the specimens had been named originally via some noticeable overall resemblance to one parent, plus a guess (e.g. by proximity) for the other. One experiment was carried out by running the program on a specimen of a species (*R. idaeus*) which was not included in the data. This showed up clearly by only scoring a maximum similarity of about 40%.

In order to improve the performance, an obvious, but tedious, approach would be to refine the descriptions of the taxa by considering more specimens. The present data was based on not more than two or three examples in each case. It would be possible to take newly determined specimens and add their descriptions to the data base as one goes along, as Rypka does with microbes (this volume, Chapter 11). There is a risk with continuous classification and identification of this kind that the effect may only be to accumulate errors, and

D

to broaden the definition of species to the extent that they become indistinguishable.

ADVANTAGES AND DISADVANTAGES

It will be convenient to compare the matching method with the operation of a diagnostic key. The matching method will always give some sort of answer, and even when the results cannot be interpreted, one may take this as evidence that the specimen cannot be identified with an existing taxon. An estimate of the quality of the result is obtained as a percentage. A restricted number of errors and omissions in the description of the specimen can be tolerated, and hybrids can be tackled with some hope of drawing a conclusion. All these are advantages over a key. An obvious disadvantage is the fact of having to complete a more detailed description, and also having to go through data preparation and have access to a computer.

OTHER WORK

All the existing publications appear to relate to pollen or microorganisms, rather than higher plants or animals. Walker *et al.* (1968) covered a wide variety of different pollen grains, and in order to do this, they divided the matching process into two levels, first in order to detect particular broad classes of characters, and then to achieve detailed matching within classes. Hansen and Cushing (1972) concerned themselves with five closely related pollen grains, and distinguished them by a matching technique which uses a probabilistic similarity measure. Campbell (1973) used a matching method like that described here on yeasts of the genus *Saccharomyces*, as part of a clustering scheme.

CONCLUSIONS

The matching method shows promise for identifying higher plants, particularly where these are closely related and the identification problem is therefore most severe. Although it requires more labour than other methods, it gives correspondingly more useful results. This kind of approach may help to make the knowledge of an expert readily available for others who wish to use it, and give greater precision to taxonomic work on difficult groups.

ACKNOWLEDGEMENTS

My thanks are particularly due to Dr S. M. Walters, for criticism and encouragement throughout, and Mrs R. R. Aitchison, for much work in specimen collection and description and in data preparation. Dr H. J. B. Birks kindly allowed us to use his clustering program.

REFERENCES

CAMPBELL, I. (1973). Computer identification of yeasts of the genus *Saccharomyces. J. gen. Microbiology* **77**, 127–135.

HALL, A. V. (1969). Group forming and discrimination with homogeneity functions. *In* "Numerical Taxonomy" (A. J. Cole, ed.). Academic Press, London and New York.

HANSEN, B. S. and CUSHING, E. J. (1973). Identification of pine pollen of late Quaternary age from the Chuska Mountains, New Mexico. *Geol. Soc. Am. Bull.* **84**, 1181–1200.

NAGY, G. (1968). State of the art in pattern recognition. *Proc. IEEE* **56**, 836–862.

SIBSON, R. (1973). SLINK: An optimally efficient algorithm for the single link cluster method. *Computer J.* **16**(1), 30–34.

SNEATH, P. H. A. and SOKAL, R. R. (1973). "Numerical Taxonomy", p. 132. Freeman, San Francisco.

WALKER, D., MILNE, P., GUPPY, J. and WILLIAMS, J. (1968). The computer assisted storage and retrieval of pollen morphological data. *Pollen et Spores* **10**(2), 251–262.

WATSON, W. C. R. (1958). "Handbook of the Rubi of Great Britain and Ireland". Cambridge University Press.

7 | Rapid Techniques for Automatic Identification

G. J. S. ROSS

Rothamsted Experimental Station, Harpenden, Hertfordshire, England

Abstract: Simple identification techniques may be sufficient for many applications, and rapid processing allows the computer user to try different approaches. The identification facilities in the author's program CLASP are described, including the construction of binary keys, finding the approximate minimum subset of variates for binary identification, identification by similarity coefficients, and the analysis of clusters to find useful identification variates.

Key Words and Phrases: identification, similarity coefficient, nearest neighbour, principal coordinates, binary key, batch key, information statistic, most typical element

INTRODUCTION

A small bird flies down in front of us and then disappears into a nearby wood. We should like to name it, but we do not have time to study it in great detail. A cow dies on a farm and the cause of death must be established. Samples of suspect food contain unidentified bacteria. These are simple examples of identification problems in which the basic feature is the comparison of attributes of *new units* with the attributes of an existing *reference set* of units.

The computer is used either directly, by presenting the new units and the reference set together, or indirectly in the sense that analysis of the reference set produces an identification scheme simple enough to use away from the computer. The conventional keys and picture books are successful wherever the computation can be done in the head.

The attributes used in identification are regarded either as fixed or as variable quantities. Identification schemes involving fixed attributes such as presence or absence of well defined features suffice for reasonably heterogeneous reference sets, but for more homogeneous material it may not be possible to find suitable

Systematics Association Special Volume No. 7, "Biological Identification with Computers", edited by R. J. Pankhurst, 1975, pp. 93–102. Academic Press, London and New York.

fixed attributes and variable quantities must be used. Even when all attributes are fixed it is preferable to allow for some degree of error in describing the units. Variable attributes may be termed "probabilistic" and fixed attributes "non-probabilistic" (Gower, this volume).

For some applications it is necessary to identify with an individual unit in the reference set, such as a species of plant, whereas for other applications it is sufficient to identify a new unit with one of a group of reference units.

It is not usually necessary to supply information on every attribute to obtain identification, and it is of interest to determine the minimum amount of information necessary. In other cases much information may be unavailable, and it is then necessary to identify as effectively as possible.

This chapter describes some of the identification aids available in the author's computer program, CLASP, which is an integrated program for classification and identification. The facilities include:

(a) Construction of binary keys attempting to minimize overall length (p. 95),

(b) finding a subset of binary attributes to achieve maximum identification (p. 96),

(c) direct identification using similarity coefficients (p. 98),

(d) analysis of groups of units to find the most informative attributes (p. 100), the most typical members of a group, and the commonest or mean values of each attribute,

(e) principal coordinate analysis, its relation to each attribute, and the assignment of coordinates to new units (p. 99).

ORGANIZATION OF THE PROGRAM

The program is concerned with the creation and manipulation of what are termed "data structures". The most important of these are as follows:

(a) The *Data Matrix*, consisting of the values of each attribute or *variate* for N_R reference units and N_E extra units. Variates may be of different types, binary, multistate or quantitative, and variate values may be missing.

(b) The *Similarity Matrix* of similarity coefficients between each pair of units of the reference set. Similarity coefficients are calculated in various ways, but all variates are equally weighted.

(c) The *Grouping Factor*, assigning units to different groups. Groups may be defined by the user or found by the clustering procedures in the program.

(d) *Principal Coordinates*, derived from the similarity matrix in such a way that pairs of similar units are represented by points close together in multi-dimensional space (Gower, 1966).

The operations of the program are concerned with reading the data, computing various structures, analysing the relationship between the structures, and printing the results in readily usable form.

Variate values of the reference set may be available in full, but when most data values are missing for the extra units a good deal of punching can be saved by specifying that all values are missing except those about to be read.

Normally the program is used several times to study a set of data fully. For example it may become clear that certain variates or units are best discarded from the analysis, or that different forms of similarity coefficient should be used.

THE CONSTRUCTION OF BINARY KEYS

When all variables are binary and assumed to be definitive binary keys, in which a sequence of decisions leads to the identification of a particular unit can be constructed for use away from the computer. Many criteria have been suggested for the construction of such keys, some of which use the relative frequencies of the units so that common units are identified more rapidly than rare ones, or take account of the relative costs of using each variate so that costly variates are not used unless they are strictly necessary.

In the program two simple procedures are available which ignore frequency of units.

1. Binary Keys of Shortest Length

The first algorithm attempts to minimize the mean number of decisions required to identify any unit. At each stage a subset of units is divided into two by selecting the variate that divides it into the most equal portions. Although this process does not necessarily lead to the best possible solution, it may not be worthwhile spending an excessive time trying to improve on the solution as it will rarely make any useful difference. Variates containing missing values may be used so long as there are no missing values in the group being split. The user should exclude any unsuitable variates, and he may select the first few variates (for example the units form natural clusters which are to be identified at the first stage in the key).

An example of the results of the algorithm is provided by Table I. Note that variate 3 is used to split the group consisting of units 3, 4 and 7.

Groups are left unsplit if no variate can be found to split them. For data in which the proportion of missing values is large this algorithm is unsatisfactory, and a method such as that devised by Gower and Barnett (1971), and described by Payne (this volume, chapter 4), may be required.

TABLE I. Binary key of shortest length for 7 units

Units	Variates					
	1	2	3	4	5	6
1	1	1	—	0	1	0
2	0	1	0	0	1	1
3	1	0	0	1	1	0
4	1	0	0	0	0	0
5	1	1	—	1	1	1
6	0	1	0	0	1	0
7	1	0	1	0	1	1

1 If variate 2 = 0 go to 2, if = 1 go to 3
2 If variate 3 = 0 go to 4, if = 1 go to 5
3 If variate 1 = 0 go to 6, if = 1 go to 7
4 If variate 4 = 0 go to 8, if = 1 go to 9
5 Identify 7
6 If variate 6 = 0 go to 10, if = 1 go to 11
7 If variate 4 = 0 go to 12, if = 1 go to 13
8 Identify 4
9 Identify 3
10 Identify 6
11 Identify 2
12 Identify 1
13 Identify 5

2. Batch Keys

The second algorithm aims to provide a key using the smallest possible number of variates simultaneously. This type of key is important for bacteriologists who may need to prepare several biochemical tests taking up to 24 h to give results, but which may be used on several unknown strains simultaneously. The algorithm computes for each variate the reduction in *Information* (Shannon, 1948) resulting from the inclusion of the variate in the batch. If the set of units is grouped into subsets of sizes $g_1, g_2, \dots g_k$, then the information in the grouping is defined as

$$\sum_{i=1}^{k} g_i \log_2(g_i)$$

and this quantity becomes zero when all groups are of size 1, which means that all units have been identified. The maximum reduction of information from the inclusion of any one variate is N, the total number of units, and this is achieved only if every group is split exactly in two, which is not in general possible unless $N = 2^r$, in which case it may be possible to identify using r variates only. In general considerably more than $1 + \log_2 N$ variates are needed, and if closely similar units are to be identified they will differ in respect of very few variates, one of which will have to be included in the batch.

The algorithm proceeds by selecting at each stage the variable that achieves the maximum reduction in information, until no further reduction is possible, which means that any groups of size greater than one cannot be further subdivided. This method does not guarantee that the minimum number of variates is being used, but the solution is usually adequate for practical purposes. The user may select the first few variates, and this is useful where some variates are particularly important, either as indicators of major subdivisions, or because they are easy to measure. An example of the algorithm is given in Table II.

TABLE II. Batch key algorithm on data of Table I

Initial information = 19·7						
Information after selecting variate	1	2	3	4	5	6
	13·6	12·8	—	13·6	15·5	12·8
After selecting variates 2 and	1		3	4	5	6
	8·8		—	6·8	10·0	6·0
After selecting variates 2, 6 and	1		3	4	5	
	2·0		—	2·0	4·0	
After selecting variates 2, 6, 1 and			3	4	5	
			2·0	0·0	0·0	

Variates selected					
2	6	1	4		Unit
0	0	1	0		4
0	0	1	1		3
0	1	1	0		7
1	0	0	0		6
1	0	1	0		1
1	1	0	0		2
1	1	1	1		5

Much computing time may be saved by calculating a table of $n \log_2 n$ from 1 to N. A list should be kept of variates that achieve no reduction in information, as these need not be tested again. The group number of each unit is stored as a binary number whose bits correspond to the values of the variates in the batch. At each selection the group numbers are multiplied by two and the new variate value is added. Group numbers are finally sorted into ascending order to provide a usable key. Variates with missing values may be used if such values are associated with units that are already identified.

Multistate variables are excluded mainly because it would complicate the algorithm to include them, although potentially a multistate variable can reduce the information by $N \log_2 k$, where k is the number of states. In practice multistate variables rarely arise in the contexts where this algorithm is useful.

IDENTIFICATION USING SIMILARITY COEFFICIENTS

A similarity coefficient is equally useful for identification as for classification. The advantages over the use of keys are that the similarity coefficient uses all the available information and does not direct the choice of variables. The disadvantages are that the identification may be incomplete or inefficient because the variates measured have little discriminatory power. However, in the presence of variability or uncertainty of establishing the value of a variate, there is much to be said for pooling the information obtained from all the variates measured.

To use the program the user must declare how each variate is to be scored in the similarity coefficient, whether as a qualitative or a quantitative variate, and whether zero matches are to be scored or ignored. Qualitative variates score 1 for a match and 0 for a mismatch, whereas quantitative variates are scored by linear or quadratic interpolation. To give quantitative variates the same weight as binary variates it is necessary to declare the range of each variate, so that the maximum value scores 0 when compared with the minimum value. Variates are omitted from a particular comparison if the value is missing in either unit.

Two levels of identification are possible: identification of single units and identification as members of a group of units. For the former, units of the reference set are ranked in decreasing order of similarity with the new unit, and the first few values are printed. For the latter, units of the reference set are already grouped, and in addition to the most similar individuals, the mean similarity coefficient with each group is computed, and the groups are ranked in decreasing order of similarity.

According to the similarities computed the user may infer:

(a) that the new unit is definitely more similar to one unit than to another;

(b) that the new unit is similar to two or more units, and there is insufficient data to distinguish them;

(c) that the new unit is a borderline case between two or more units;

(d) that the new unit bears little resemblance to any of the reference units.

Table III exemplifies the use of similarity coefficients where data based on the sighting of small land birds are compared with a data matrix of field marks of British birds taken from Fitter and Richardson (1966). Such characters as overall size, tail ratio, major colour, colours of various parts of the body, distinguishing marks, flight features, habitat, time of year and locality were coded into 60 variables, but for any one sighting it was assumed that only about 10 of these could be coded. Note that the method failed to identify the Willow Tit, which is in practice difficult to distinguish from the Marsh Tit, but otherwise the identifications were correct.

A similar method has been used with success by Pankhurst (this volume).

TABLE III. Identification of birds using 8–10 out of 60 possible variates

Percent similarity of five nearest neighbours in the reference set					
Unit A (Wheatear)		Unit B (Marsh Tit)		Unit C (Wren?)	
Wheatear	100	Willow Tit	100	Wren	85·7
Woodlark	75	Marsh Tit	100	Meadow pipit	71·4
Meadow pipit	75	Wren	87·5	Hedge sparrow	71·4
Tree pipit	62·5	Woodlark	75	Chaffinch	71·4
Nuthatch	62·5	Tree creeper	75	House sparrow	71·4

In Unit C the size of the bird was incorrectly described.

1. Use of Principal Coordinate Analysis

Principal Coordinate Analysis was developed by Gower (1966) and provides a graphical representation of the units based on a similarity coefficient or distance measure. Principal component analysis is a special case applicable when all variates are quantitative (or binary), but Principal Coordinate Analysis handles data of mixed type and does not object to missing values.

For the purpose of identification Gower (1968) developed a technique for adding points to a principal coordinate analysis using the similarity coefficients between the new unit and each of the reference units. Its use is supplementary to the method described in the previous section, and it is not advisable to identify points with their closest neighbours in any two-dimensional plots unless this plot summarizes most of the similarity matrix. However it is a useful method of distinguishing between "intermediates" and "outliers", especially when identifying members of a group.

2. Identification as a Means of Enlarging a Classification

The identification of a unit as a member of a group effectively enlarges the group. This leads to an approximate but economical method of cluster analysis in which groups are formed by allocating units to their nearest neighbours in the reference set, without computing the full similarity matrix. This technique is suitable for medical and sociological data in which the number of units is very large but exact matching is not expected. The method has been refined by allowing a measure of overlap so that a minimum spanning tree may be computed (Ross, 1969).

MISCELLANEOUS IDENTIFICATION AIDS

Various facilities provided by the program may be regarded as aids to identification, although their use is guided by the program user.

1. Group–Variate Interactions

If the units have been grouped then the values of each variate may be analysed to see whether the variate is likely to be useful in a key to the groups. For binary or multistate qualitative variates a value of χ^2 is computed for each group as if the frequencies of each value were being compared with theoretical frequencies, although in practice the marginal frequencies of the table are used. This statistic, although no "significance" level may be attached to it, is very effective at highlighting the groups whose value distribution differs substantially from the marginal distribution.

For quantitative variates a t-statistic is computed, to highlight groups for which the mean score is substantially higher or lower than the overall mean.

These procedures would be statistically valid if the groups were chosen *a priori* without reference to the data variates under study. Groups formed on the basis of the data are expected to be "significantly" different in respect of several variates. Nonetheless in practice the statistics are extremely useful when there are many variates.

The analysis for a particular variate is given in Table IV.

Variates may be selected from this analysis to control the initial stages of the key-forming algorithms described in the section "The Construction of Binary Keys".

A similar analysis is provided to assist in the interpretation of principal coordinates. A variance ratio statistic is computed for each coordinate and each variate.

TABLE IV. Analysis of frequencies of variate values for groups found by cluster analysis

Group	Variate values						Total	"Chi-squared"
	1	2	3	4	5	6		
1	6	0	2	0	0	0	8	21·1
2	0	1	5	2	0	0	8	3·9
3	0	5	0	1	0	0	6	16·5
4	0	2	2	0	1	1	6	15·3
5	0	0	10	1	0	0	11	8·1
6	2	2	6	1	0	0	11	0·5
Total	8	10	25	5	1	1	50	

The values of "chi-squared" show that groups 1, 3 and 4 are markedly different from average in respect of the values of this variate.

2. *Most Typical Units of a Group*

If the units in each group are ranked in order of mean similarity with other members of the group, the units with highest mean similarity may be regarded as "type" units, a concept familiar to microbiologists. These type units may then be used to form the basis of a reference set for future identification by the nearest neighbour method described in the previous section.

A further method of finding the centre of a group is to construct a fictitious "centre" unit from the data values of each variate in the following way: for qualitative variates assign the value of any variate that occurs in more than 61% of units in a group (this value is arbitrarily chosen so that nearly equally divided binary variates will be excluded) and assign the missing value if no one value predominates; for quantitative variates assign the mean value. This method is useful in medical and sociological applications where a fictitious "typical case" may be described. A similar concept is the "class predictor" described by Gower (this volume). In the program the group centres so formed may be arranged to replace the original data matrix so that new units for identification may be read and matched by the nearest neighbour method. The advantage of the fictitious "centre" over the "type" unit is that the latter may have some properties irrelevant to the rest of the group which are best ignored in identification.

DISCUSSION

The facilities described above provide a useful basis for identification without excessive computation. However, they are based on descriptive statistics only, and many problems such as the distributions of variates within groups, the non-independence of variates, weighting of variates by cost and prior probabilities of occurrence of units, are ignored. To some extent absence of knowledge about the distribution of variates is offset by the use of a large number of variates, for if two different units take the same values for several variates there is still the chance that there will be further variates in which their values will differ.

The program CLASP is written in ASA Standard FORTRAN and is used on the ICL 4.70 computer at Rothamsted Experimental Station. Some of the facilities discussed are also available in the Rothamsted general statistical program GENSTAT (Nelder, 1973).

ACKNOWLEDGEMENT

I thank Mr F. B. Lauckner and Miss Diana Hawkins for assistance in writing the program.

REFERENCES

FITTER, R. S. R. and RICHARDSON, R. A. (1966). "Collins' Pocket Guide to British Birds" (Revised edition). Collins, London.

GOWER, J. C. (1966). Some distance properties of latent root and vector methods used in multivariate analysis. Biometrika **53**, 325–338.

GOWER, J. C. (1968). Adding a point to vector diagrams in multivariate analysis. Biometrika **55**, 582–585.

GOWER, J. C. and BARNETT, J. A. (1971). Selecting tests in diagnostic keys with unknown responses. *Nature, Lond.* **232**, 491–493.

NELDER, J. A. (1973). "GENSTAT Reference Manual". University of Edinburgh Scientific and Social Sciences Program Library, Report No. 3.

ROSS, G. J. S. (1969). Classification techniques for large sets of data. *In* "Numerical Taxonomy". (A. J. Cole, ed.). Academic Press, London and New York.

SHANNON, C. E. (1948). A mathematical theory of communications. *Bell Systems Techn. J.* **27**, 379–423; 623–656.

8 | Methods Used in a Program for Computer-aided Identification of Bacteria

W. R. WILLCOX and S. P. LAPAGE

National Collection of Type Cultures,
Central Public Health Laboratory, London, England

Abstract: The methods incorporated in the program used in a computer-aided identification service for bacteria are described. The identification method is based on Bayes' theorem and allows for dependent and multistate tests and missing data in the probability matrix. If a definite identification is not possible on the initial test results a separate procedure selects the best tests to continue the identification.

The use of weights of evidence to assess the discrimination between taxa provided by a probability matrix is described.

Key Words and Phrases: identification, test selection, Bayes' theorem, weights of evidence, bacteria

INTRODUCTION

A computer program, incorporating the methods described in this paper, is used in an identification service for Gram-negative rod-shaped bacteria. To make use of the service a laboratory completes an application form and sends it to us together with a culture of the strain of bacteria to be identified. Bacteria are identified by the results of biochemical tests and the application form lists the 90 available tests for which a matrix of probabilities is held. The sending laboratory will have carried out a number of these tests, usually about 20, and enters the results obtained on the form. At our laboratory a paper tape is punched from the application form and the results are processed by the program which is operated on a commercial computing service to which access is obtained by a Teletype terminal. For each strain the program prints a report which contains reference information, the test results used, and the results of the identification calculation (Figs 1 and 2 give examples of such reports). The report may indicate that the strain cannot be identified at this stage and list a number of

Systematics Association Special Volume No. 7, "Biological Identification with Computers", edited by R. J. Pankhurst, 1975, pp. 103–119. Academic Press, London and New York.

```
FOR:ANYWHERE PHL                         OUR REF:9999/73 RUN W1
DATE:04/09/73                               (M511 LAB.0235)

YOUR REF:1234          SMITH,JOHN

YOUR RESULTS USED
 - MOTILITY 37      50        - ADONITOL PWS     1
 + GROWTH 37        99        + ARABINOSE PWS..99
 + GROWTH RT        99        - DULCITOL PWS     1
 - PIGMENT          75        - INOSITOL PWS     5
 + MACCONKEY        99        + LACTOSE PWS     30
 + CATALASE         99        + MANNITOL PWS    99
 - OXIDASE           1        + SUCROSE PWS     90
 + INDOLE            1
 - MR 37            99        + GEL 1-28/PLATE 95
 - VP 37             1
 + SIMMONS CITR     70
 - UREASE            1
 + KCN              15
 - H2S PAPER         1
 - H2S TSI           1
 - GLUCONATE        30
 - MALONATE         99
 + ONPG             99
 - PPA               1
 - ARGININE          1
 - LYSINE            1
 - ORNITHINE         1
 + GLUCOSE PWS      99
 - GAS GLUCOSE       5

 32 TESTS DONE
NOT IDENTIFIED, TESTS SELECTED(KEY=   90)

    TEST       VALUE IN SET
MOTILITY RT       25
RHAMNOSE PWS      20
STARCH PWS        15
RAFFINOSE PWS      8
SORBITOL PWS       6
CELLOBIOSE PWS     3
MR RT              3
HUGH & LEIFSON     1
SELENITE 0.4       1
MALTOSE PWS        1
SALICIN PWS        1
VP RT              1
  END OF SET,VALUE=   85

XYLOSE PWS        18
GELATIN 1-5       18
NITRATE            9
  END OF SET,VALUE=   45


        GROUP                         SCORE
ERWINIA HERBICOLA                   0.769576
AEROMONAS FORMICANS                 0.152537
PASTEURELLA MULTOCIDA               0.020895
AEROMONAS HYDROPHILA                0.017652
KLEBSIELLA AEROGENES & K.OXYTOCA    0.015881
E.COLI(BIOCHEMICALLY ATYPICAL)      0.007896
KLEBSIELLA OZAENAE                  0.006612
VIBRIO CHOLERAE                     0.002217
ACINETOBACTER ANITRATUS             0.001814
CITROBACTER KOSERI                  0.001748
```

FIG. 1. A report produced by the identification program. This "run" used the sender's test results. The strain does not identify but additional tests are suggested.

```
FOR:ANYWHERE PHL                    OUR REF:9999/73 RUN R2
DATE:17/09/73                            (M511 LAB.0235)

YOUR REF:1234         SMITH, JOHN

COMBINED RESULTS USED
  - MOTILITY 37    99*        - ADONITOL PWS     1
  - MOTILITY RT    99<        + ARABINOSE PWS    85
  + GROWTH 37      99*        + CELLOBIOSE PWS   85<
  + GROWTH RT      99*        - DULCITOL PWS     1
  - PIGMENT         1*        - INOSITOL PWS     1
  + MACCONKEY      99         + LACTOSE PWS      50
  + CATALASE       99*        + MALTOSE PWS      99<
  + OXIDASE        99#        + MANNITOL PWS     99
  + H&L FERM       99<        - RAFFINOSE PWS    1<
  + INDOLE         85*        - RHAMNOSE PWS     1<
  + MR 37          99#        + SALICIN PWS      85<
  + MR RT          99<        - SORBITOL PWS     10<
  - VP 37           1*        + SUCROSE PWS      99
  - VP RT           1<        + STARCH PWS       99<
  + SIMMONS CITR   70
  - UREASE          1         + SELENITE 0.4     95<
  + KCN            85*        + GEL 1-28/PLATE   99
  - H2S PAPER      80*
  - H2S TSI         1
  - GLUCONATE       1
  - MALONATE        1*
  + ONPG           99
  - PPA             1*
  + ARGININE       99#
  - LYSINE          1
  - ORNITHINE       1
  + GLUCOSE PWS    99
  - GAS GLUCOSE     1

  44 TESTS DONE
IDENTIFIED

AEROMONAS FORMICANS
 UNUSUAL RESULTS IN:
MOTILITY 37
MOTILITY RT

        GROUP                 SCORE
AEROMONAS FORMICANS         0.999988
AEROMONAS HYDROPHILA        0.000012

  *=OUR RESULT AGREES WITH YOURS
  <=OUR RESULT ONLY
  #=OUR RESULT(CL) DIFFERS FROM YOURS(SL),OURS USED
                  CL   SL
      OXIDASE      +   - @
      MR 37        +   - @
      ARGININE     +   - @
      @=UNEXPECTED RESULT FOR THIS ORGANISM
```

FIG. 2. A report produced by the identification program. This "run" used the sender's test results combined with results obtained in our laboratory. The strain identifies but unusual test results are reported.

selected tests with which to continue the identification (Fig. 1); or it may suggest that the strain can be identified as a member of one of the 90 available taxa (Fig. 2); or it may indicate that the strain cannot be identified and that none of the remaining tests is likely to further the identification. In each case the most likely taxa are listed with their "identification scores" and beside each test result is printed the estimated probability that a member of the highest scoring taxon will give a positive result in that test (the identification calculation is based on a matrix of these probabilities, see below). If the strain has identified but has given unusual results in some of the tests a list of these tests is printed (Fig. 2); these are the tests in which the strain gave a positive result and the probability of a positive result for the taxon to which the strain identified is 0·01, or a negative result when the probability of a positive result is 0·99. The report is examined by a member of our laboratory and if necessary additional tests are carried out in this laboratory. The results of these tests are entered on the application form and a further computer "run" carried out. On this and subsequent runs the results obtained in this laboratory are combined with the sender's results. If the results of a particular test conflict, our result is usually taken because we use standardized methods and we are able, if necessary, to repeat any test. The report for such a run (Fig. 2) indicates beside each test result which laboratory obtained the result and lists the tests in which conflicting results were obtained. The cycle of testing, computer runs and examination of the computer reports continues until the bacteriologist is satisfied with the identity of the strain or considers that it cannot be identified. The final computer report, with an additional typed report or explanatory letter if necessary, is returned to the sending laboratory.

The system involves continual assessment of the computer-produced reports by the bacteriologist responsible for the identification. The bacteriologist may occasionally decide that the identification suggested by the computer is not justified, often because of the unusual results indicated on the report. Alternatively the bacteriologist may be able to identify a strain which the computer program was unable to identify. Such identifications will usually be to one of the likely taxa suggested by the report and may make use of tests not available to the program. The bacteriologist may also choose to carry out tests other than those selected by the computer; in particular tests may be repeated if the report suggests that the results originally obtained are unusual for the most likely taxon. Thus, although the reports produced by the program assist identification, the automatic system must be monitored by a bacteriologist.

The identification program was tested by applying it to 1079 reference strains, aberrant as well as typical strains, whose identity had already been well established (Bascomb *et al.*, 1973). Of these strains, 89% were correctly identified, a single strain was incorrectly identified and the remainder were not identified by the program. The identification service was operated in an experimental form from 1966 to 1971 (Lapage *et al.*, 1973) and has been operated as described above since 1972.

IDENTIFICATION METHOD

The identification method is based on Bayes' theorem which has been widely used in medical diagnosis (Ledley and Lusted, 1959). In terms of the identification of bacteria Bayes' theorem states

$$P(t_i \mid R) = \frac{P(t_i)P(R \mid t_i)}{\sum\limits_i P(t_i)P(R \mid t_i)}, \qquad (1)$$

where $P(t_i \mid R)$ is the probability that an organism giving a set of test results R is a member of taxon t_i; $P(t_i)$ is the probability that an organism is a member of taxon t_i before considering any test results, the prior probability of the taxon; and $P(R \mid t_i)$ is the probability that a member of taxon t_i will give test results R. The prior probabilities are the expected frequencies of incidence of the taxa among the strains submitted for identification. They were set equal for all taxa at the outset because there was little indication of how the taxa would be represented in the strains sent to such a service, and have been left equal to avoid bias against the identification of strains of the rarer taxa. For equal prior probabilities equation (1) becomes

$$P(t_i \mid R) = \frac{P(R \mid t_i)}{\sum\limits_i P(R \mid t_i)}. \qquad (2)$$

To estimate the probability of obtaining a particular result pattern, R, given a taxon t_i it was assumed that the individual test results which make up the pattern are independent in each taxon. The probability of a number of independent results is the product of their individual probabilities so

$$P(R \mid t_i) = P(r_1, r_2 \dots r_n \mid t_i) = P(r_1 \mid t_i)P(r_2 \mid t_i) \dots P(r_n \mid t_i), \qquad (3)$$

where $r_1 \dots r_n$ are the individual test results and $P(r_1 \mid t_i)$ etc. are their probabilities. The validity of this assumption, which again has been widely adopted in medical diagnosis (Anderson and Boyle, 1968) is discussed below.

Most of the tests used are of the two-state (+ or −) type and for such tests the probability of a positive result is one minus the probability of a negative result, so a table or matrix giving the probability of a positive result for each of the taxa in each of the tests is all the information required to calculate by equations (3) and (2) the probabilities of all the taxa, $P(t_i \mid R)$, for any set of test results, R. These probabilities are referred to as "identification scores" on the computer-produced reports and if the score of one of the taxa exceeds 0·999 an identification to this taxon is suggested.

The assumption that individual test results are independent, except for a few straightforward cases described below, is a major theoretical objection to the method. Even without statistical examination, it is known on bacteriological grounds that several of the tests are likely to be correlated within some or all of the taxa, though correlated tests were avoided as far as possible in the initial choice of the tests to be used. Methods are available which do not depend on this assumption (e.g. Dickey, 1968) but an advantage of the present method is the simple form of the basic data, the probability matrix. This makes it possible to combine information on the behaviour of the taxa in the tests from different sources in the literature with records of sample strains, the final values being assessments, to some extent subjective, of the available data (Lapage *et al.*, 1973). The matrix is also easily modified to allow for new findings though it was not automatically adjusted with each strain identified (Lapage *et al.*, 1973).

Some problems which arose when the identification method was applied, and the procedures adopted to overcome them, will now be described. The procedures have been described more fully by Willcox *et al.* (1973).

1. *Limits on matrix entries*

If, for a particular taxon, all the organisms which had been examined had given positive (or negative) results in a particular test, and if the appropriate matrix element was given a value of one (or zero), then an organism giving a test result never before observed in a taxon could never be identified as a member of that taxon whatever its results in the remaining tests (if any $P(r_j \mid t_i)=0$ in equation (3) then $P(R \mid t_i)=0$ and $P(t_i \mid R)=0$ in equation (2)). This is undesirable since any previous sampling cannot have discovered all possible test result patterns for a given taxon, and furthermore the result in question might have been due to an error in testing or transcription.

The problem was avoided by setting upper and lower limits of 0·99 and 0·01 to matrix entries. The initially arbitrary choice of these limits was later justified to some extent by the data error approach described below.

2. *Unknown matrix entries*

A test may need to be included in the matrix because it is particularly useful in discriminating between some of the taxa, although data are not available on the results of strains of some of the other taxa. This situation has not been very frequent in our project and we have been able to complete the data by testing sample strains. However, the method described here which allows for unknown values in the matrix enables such tests to be used while further data are being collected.

Suppose the value required to obtain the probability of an individual test result given a taxon, $P(r_j \mid t_i)$ of equation (3) is unknown. If the test result is positive a value of 0·99 is assumed, if negative a value of 0·01, in either case $P(r_j \mid t_i) = 0·99$ its highest possible value. This procedure is applied to all the unknown matrix entries encountered for a given set of tests and if the identification score of one of the taxa exceeds 0·999 a tentative identification to that taxon is made. Then if there are no unknown matrix entries for the tentatively identified taxon in the tests considered the identification is accepted because the values assumed for the unknown entries of the other taxa were those most favourable to these taxa and thus the least favourable to the identifying taxon. If, on the other hand, unknown matrix entries for the tentatively identified taxon had been encountered the identification could not be accepted because values favourable to the identifying taxon have been assumed for these entries. The identification scores are recalculated without assuming values favourable to the tentatively identified taxon, a process termed "rescoring". If the same taxon identifies after rescoring the identification is accepted, if not, test selection is attempted. Two strategies have been used for rescoring. In the "lenient strategy" the test results for which the matrix entries of the tentatively identified taxon are unknown are ignored. In the "stringent strategy" values are assumed for these entries which are the least favourable, subject to the limits on matrix entries, to the tentatively identified taxon. In either case unknown matrix entries for taxa other than the tentatively identified taxon are treated as before; the most favourable values are assumed.

A disadvantage of the lenient strategy is that some of the discrimination between the lower scoring taxa may be lost since some of the test results are ignored. In the full trial of the identification method (Bascomb *et al.*, 1973) there were no unknown matrix entries and rescoring was only required in the treatment of linked tests (see below).

3. Linked tests

Some of the tests in the matrix are known to be linked and for these tests a modification to the basic method is needed. The results of a set of linked tests are governed by a simple logical relationship valid for all taxa, e.g. if a strain does not grow at 37°C the result for motility at 37°C must be negative. Formulae for obtaining the probabilities of combinations of results of such linked tests are easily derived and have been included in the computer program (Willcox *et al.*, 1973).

For example, if test 1 (e.g. growth at 37°C) is linked to test 2 (e.g. motility at 37°C) such that a negative result in test 1 implies a negative result in test 2 then

$$P(+_1, +_2 \mid t_i) = P(+_1 \mid t_i) \times \frac{P(+_2 \mid t_i)}{P(+_1 \mid t_i)} \tag{4}$$

$$P(+_1, -_2 \mid t_i) = P(+_1 \mid t_i) \times \left[1 - \frac{P(+_2 \mid t_i)}{P(+_1 \mid t_i)} \right] \tag{5}$$

$$P(-_1, +_2 \mid t_i) = P(-_1 \mid t_i) \times 0 \tag{6}$$

$$P(-_1, -_2 \mid t_i) = P(-_1 \mid t_i) \times 1 \tag{7}$$

where, for example, $P(+_1, +_2 \mid t_i)$ is the probability of positive results in both test 1 and test 2 given taxon t_i; and $P(+_1 \mid t_i)$ is the probability of a positive result in test 1 obtained from the matrix in the usual way. The results $-_1, +_2$ (i.e. the strain does not grow at 37°C but is motile at 37°C) given a zero probability by (6) are detected as "impossible" by an earlier editing procedure which rejects the test results and prints an error message.

Consider equation (5) above, if $P(+_1 \mid t_i) = P(+_2 \mid t_i)$ then $P(+_1, -_2) = 0$ which is unacceptable for the reasons given under limits on matrix entries above. Such probabilities occur quite frequently in the present matrix, in all cases $P(+_1 \mid t_i) = P(+_2 \mid t_i) = 0.99$ or 0.01. Reasonable results can be obtained by recognizing that, because of the limits on matrix entries, an entry of 0.99 represents a probability somewhere between 1 and 0.99 and an entry of 0.01 a probability between 0 and 0.01. Then, using the procedure for unknown matrix entries, the most favourable values are first assumed; for $P(+_1 \mid t_i)$ and $P(+_2 \mid t_i)$ entered as 0.99 assume $P(+_1 \mid t_i) = 1$, $P(+_2 \mid t_i) = 0.99$ giving $P(+_1, -_2 \mid t_i) = 1 \times 0.01$; and for $P(+_1 \mid t_i)$ and $P(+_2 \mid t_i)$ entered as 0.01 assume $P(+_1 \mid t_i) = 0.01$, $P(+_2 \mid t_i) = 0$ giving $P(+_1, -_2 \mid t_i) = 0.01 \times 1$. Because values have been assumed for matrix entries rescoring will sometimes be necessary but the lenient strategy is not applicable here because one of the results is known to be unexpected for the tentatively identified taxon (e.g. $+_1$ where $P(+_1 \mid t_i) = 0.01$). The stringent strategy is not immediately applicable because the least favourable assumptions give $P(+_2) = P(+_1)$ and $P(+_1, -_2) = 0$ again. The data errors approach, below, suggests a straightforward procedure for rescoring these test results.

4. Multistate tests

As mentioned above most of the tests used are two-state and the probability of a negative result for a given taxon in a given test is not stored in the computer but is calculated as one minus the probability of a positive result. In general terms, for an n-state test, either $n - 1$ probabilities can be stored in the computer and the probability of the nth result calculated as required; or n probabilities can be held and referred to directly. For two-state tests the first method halves the amount of computer store required, for multistate $(n>2)$ tests the saving in store is proportionately less and more calculation is needed to obtain the nth probability. The first method is used in the present program for two-state tests but the second approach is used for the few multistate tests by the following procedure: an n-state test is represented in the matrix by n components treated as individual tests, the matrix entry for each component being the probability that the equivalent result will be obtained in the test for the given taxon. Any result of the test is presented as a positive result in the appropriate component test, the other component tests being scored "not done" and thus ignored.

5. Data errors approach

As mentioned above, it was necessary to set upper and lower limits on matrix entries to prevent the exclusion of a taxon because of a single unexpected test result. Initially the only basis for the values chosen, 0.99 and 0.01, was a subjective assessment by a bacteriologist of the weight to be attached to unexpected test results. The approach described here follows the reasoning that such unexpected results may be due to errors in testing.

The results of bacteriological tests are subject to experimental errors. We assume that any test result has a probability P_m of being incorrect and assume a value of 0.01 for this probability; the validity of these assumptions is considered below. Then if $P(+)$ is the "true" probability of a positive result for a given taxon in a given test and $P_a(+)$ and $P_a(-)$ are the equivalent apparent probabilities of positive and negative results allowing for errors in test results,

$$P_a(+)=P(+)(1 - P_m) + (1 - P(+))P_m$$

for $P_m = 0.01$

$$P_a(+) = 0.98P(+) + 0.01 \tag{8}$$

and similarly

$$P_a(-) = 0.99 - 0.98P(+). \tag{9}$$

If the formulae (8) and (9) were used in the identification program and $P(+)$ obtained from the matrix in the usual way there would be no need to set limits on matrix entries as $P_a(+) = 0.01$ for $P(+) = 0$ and $P_a(-) = 0.01$ for $P(+) = 1$. The matrix entries are taken as estimates of the true probabilities, the results on which they are based are assumed to be free of errors as any apparently aberrant results in this data will have been confirmed by repetition of tests. In practice there are difficulties in the application of the data errors approach (see below) and the formulae were not used directly in the program, but it is easily seen that the effect of the formulae is very similar to that of the original method of setting limits on matrix entries. Table I shows that the matrix entries and apparent probabilities differ only slightly over most of the range of probabilities and the greatest difference, a factor of about two at $P(+)=0.01$ is negligible in the present application considering the approximate nature of the matrix entries as estimates of $P(+)$.

TABLE I. The values entered in the matrix (applying upper and lower limits of 0.99 and 0.01) and the apparent probabilities (allowing for an 0.01 probability of an error in testing) for various probabilities of a positive result for a given taxon in a given test

True probability	Matrix entry	Apparent probability
1	}0.99	0.99
0.99		0.9802
0.5	0.5	0.5
0.1	0.1	0.108
0.05	0.05	0.059
0.01	}0.01	0.0198
0		0.01

Table I also shows that the limits on matrix entries imply that for entries of 0.99 and 0.01 the true probability is only known to lie within a range of values. This was useful when the matrix was compiled since the difficulties in estimating probabilities near to one and zero were avoided. To allow for the range of possible values the approach described above for unknown matrix entries is followed. The most favourable values are first assumed for all taxa, e.g. for a positive result and for a matrix entry of 0.99, assume $P(+)=1$ giving $P_a(+)=0.99$, while for a matrix entry of 0.01, assume $P(+)=0.01$ giving $P_a(+)=0.0198$. If a taxon now identifies the calculation must be repeated without assuming favourable values for this taxon. The lenient strategy of rescoring described above is not appropriate in this case because the probabilities in question are not completely unknown but are known to lie within limits. Instead, for the tentatively identified taxon, the least favourable values are assumed, e.g. for a positive result: for a matrix entry of 0·99, assume $P(+)=0·99$ giving $P_a(+)=0·9802$; for an entry of 0·01 assume $P(+)=0$ giving $P_a(+)=$ 0·01. The difference for an entry of 0·99 is clearly negligible and the difference for an entry of 0·01, a factor of about two, can be ignored since such results will be "unusual" for the tentatively identified taxon and, if any such results should occur, they will be listed on the report to be assessed by the bacteriologist. Thus rescoring is not in fact required, in agreement with the original method which only specified rescoring for completely unknown matrix entries.

In a similar way the approach can be applied to linked tests. The apparent probability for each pattern of test results, $P_a(R)$, is found in terms of $P(+_1)$, $P(+_2)$ etc., i.e. the probabilities of positive results in the individual tests. The effects of 0·99 and 0·01 matrix entries are considered by finding maximum and minimum apparent probabilities, $P_a(R)_{max}$ and $P_a(R)_{min}$, for various situations involving such entries. When this is done for the linked tests occurring in the present matrix, it is found that: (a) the previously developed method always gives a reasonable approximation, within a factor of 4, to $P_a(R)$ and where appropriate $P_a(R)_{max}$; (b) the difference between $P_a(R)_{max}$ and $P_a(R)_{min}$ is always negligible except when an unusual result is found (positive with matrix entry of 0·01 or negative with a matrix entry of 0·99); (c) for cases where $P_a(R)_{max}$ and $P_a(R)_{min}$ differ a reasonable approximation to $P_a(R)_{min}$ is given by $P_a(R') \times (0·01)^n$ where R' is the pattern of results excluding unexpected results and n is the number of such results. Thus the only addition to the original method suggested by the data errors approach is a simple method for rescoring unexpected results in linked tests.

The difficulty in the application of the data errors approach lies in the estimation of the error rate P_m. The error rate must allow for mistakes in testing, recording and transcription. Furthermore, recent studies on the reproducibility of bacteriological tests (e.g. Sneath and Johnson, 1972; Snell and Lapage, 1973) have found that repeated careful examination of the same strains in one laboratory shows a variation in test results. This variation was represented in terms of calculated error rates by Sneath and Johnson (1972) who found rates of between 0 and 14% for the 63 tests they studied with an average rate of 2%. Snell and Lapage (1973) found that 5% of their test results varied with two or three replications which, using the formulae of Sneath and Johnson (1972), also represents a calculated error rate of about 2%. We use test results obtained in other laboratories and differences in materials and test methods between laboratories are an additional source of variation for these results. Analysis of the results of tests carried out by laboratories which sent us strains and repeated in our laboratory showed differences in about 8% of test results (Lapage *et al.*, 1973) and assuming an error rate of 2% within laboratories this suggests an error rate of about 6% between laboratories. These results suggest that a rigorous application of the data errors approach should use different error rates for different tests and different rates for results obtained in laboratories other than the laboratory which constructed the matrix.

TEST SELECTION

The program only suggests a definite identification if the identification score of one of the taxa exceeds 0·999, otherwise a separate procedure selects a set of tests with which to continue.

TABLE II. Entries in a subsidiary matrix for the component tests of a 3-state test. (a) Examples of possible patterns. (b) The patterns as entered to allow the evaluation of the multistate test as a whole; once a + has been entered remaining entries are set to 0

(a)	Component Tests			(b)	Component Tests		
	a	b	c		a	b	c
1	+	−	−	1	+	0	0
2	−	+	−	2	−	+	0
3	−	−	+	3	−	−	+
Taxa 4	−	−	−	Taxa 4	−	−	−
5	−	0	−	5	−	0	−
6	0	0	0	6	0	0	0

A set of likely taxa is first found by taking taxa in order of their scores until the sum of the scores exceeds 0·999. The tests available for selection are defined by excluding those tests which have already been carried out, and any tests whose results can be predicted from results already known (e.g. motility at 37°C would be excluded if growth at 37°C was known to be negative). A subsidiary matrix of likely taxa by available tests is then formed by entering + for probabilities in the original matrix $\geqslant 0.85$, entering − for probabilities $\leqslant 0.15$, and entering 0 (variable) for other probabilities.

Tests are selected from the subsidiary matrix by a method based on that described by Gyllenberg (1963) and developed by Rypka *et al.* (1967). A test is considered to separate a pair of taxa if it has the entry + in one taxon and − in the other. Tests are selected one at a time, at each stage the test added to the list of selected tests is the one which separates the most pairs of taxa not separated by at least two of the tests already selected. When all pairs of taxa are separated by at least two tests, or when none of the remaining tests can provide any further separation the set of tests is complete and if any tests remain the selection of a further set is started.

Multistate tests are allowed for by entering the component tests (see Multistate tests, above) into the subsidiary matrix in a special way (see Table II). The resulting entries are such that at most, one of the component tests can separate a given pair of taxa and the number of separations made by the multistate test as a whole is the sum of the separations made by the component tests.

DISCRIMINATION BETWEEN TAXA

Trials of the identification program showed the importance of an adequate matrix. Improvements to the matrix as additional information was collected increased the computer identification rate for reference strains from 81 to 91% for fermentative bacteria and from 43 to 82% for non-fermentative bacteria (Lapage *et al.*, 1973). (Fermentative and non-fermentative groups, based on Hugh and Leifson's test, were considered separately because it was known that several of the tests in this matrix were not suitable for the identification of non-fermentative bacteria.)

It would be useful to ascertain, without examining a large number of strains, whether or not a particular matrix is likely to give an acceptable identification rate. A detailed analysis of the identification of reference strains (Bascomb *et al.*, 1973) suggested that two factors are important.

Firstly, as might be expected, the matrix figures for each taxon should accurately reflect the test results given by strains of that taxon. The strains were analysed by the number of results in which they differed from the matrix; a result was considered to differ from the matrix if the result was positive and the matrix entry for the taxon to which the strain was known to belong was 0.01, or the result was negative and the matrix entry was 0.99. The identification rate fell with the number of such differences, from 95% for strains with no differences to 30% for strains with four or more differences.

The second factor is the discrimination provided by the matrix; the matrix entries for a particular taxon must be sufficiently different from those of all the other taxa before strains of that taxon can be identified. Bascomb *et al.* (1973) measured the discrimination between two taxa by counting the number of tests which had matrix entries of 0.99 in one taxon and 0.01 in the other and took the minimum number of such tests for a given taxon (i.e. the number of tests which separated the taxon from its "nearest neighbour") as an indication of how well that taxon was separated from the taxa as a whole. The identification rate for strains of taxa with two or more separating tests was significantly higher than that for strains of taxa with less than two separating tests. None of the non-fermentative taxa had more than one separating test and the identification rate for these taxa did not differ significantly from the rate for those fermentative taxa with one or less separating tests, which suggested that the lower identification rate for non-fermentative bacteria was due to the lack of tests which discriminated between these taxa. Further tests have now been added to the matrix to improve the discrimination of these taxa.

The above method of measuring the discrimination between two taxa by considering only 0·99 and 0·01 matrix entries can obviously only give an approximate indication of the results of the identification method applied to actual strains. A more accurate approach based on "weights of evidence" (Good, 1950) has proved useful. The weight of evidence in favour of taxon t_i provided by test results R is defined as

$$W_i = \log \left[\frac{P(t_i \mid R)}{1 - P(t_i \mid R)} \right] \tag{10}$$

where $P(t_i \mid R)$ is the probability of taxon t_i given results R. For the particular case of only two taxa, t_1 and t_2,

$$W_1 = \log \left[\frac{P(t_1 \mid R)}{P(t_2 \mid R)} \right]$$

$$= \log \left[\frac{P(R \mid t_1)}{P(R \mid t_2)} \right]$$

and if the result pattern R is made up of independent individual results r_1, r_2, r_3 etc.

$$W_1 = \log \left[\frac{P(r_1 \mid t_1)}{P(r_1 \mid t_2)} \right] + \log \left[\frac{P(r_2 \mid t_1)}{P(r_2 \mid t_2)} \right] + \log \left[\frac{P(r_3 \mid t_1)}{P(r_3 \mid t_2)} \right] + \dots .$$

Thus the total weight is the sum of the weights provided by the individual test results and as the individual weights can be found from the matrix figures, and as the total weight is related to the identification scores, $P(t_i \mid R)$, by equation (10), the outcome of the identification scores calculation for two taxa can be readily determined for any pattern of test results. By investigation of a number of known or expected test result patterns the discrimination between the two taxa can be established. The weights of evidence calculation still does not predict exactly the results of the identification program because it considers only pairs of taxa rather than all the taxa together. In practice, however, it was found that for many of the strains which did not identify only two taxa had appreciable identification scores. If such pairs of taxa can be discovered by preliminary trials or are known from previous experience the weights of evidence method can be applied.

As an example the discrimination between *Pseudomonas acidovorans* and *P. testosteroni* is investigated. Table III gives the only tests in the matrix in which

TABLE III. Example application of weights of evidence to *Pseudomonas acidovorans* and *P. testosteroni*

Test	Matrix figures		Weights of evidence		"Swing"
	P. acidovorans	*P. testosteroni*	+ result	− result	
glucose ASS	0·15	0·01	1·18	− 0·07	1·25
glycerol ASS	0·99	0·01	2·00	− 2·00	4·00
mannitol ASS	0·85	0·01	1·93	− 0·82	2·75
ethanol ASS	0·35	0·01	1·54	− 0·18	1·72
fructose ASS	0·70	0·01	1·85	− 0·52	2·37

Typical strain of *P. acidovorans*		Typical strain of *P. testosteroni*	
Result	Weight	Result	Weight
glucose ASS −	− 0·07	glucose ASS −	− 0·07
glycerol ASS +	2·00	glycerol ASS −	− 2·00
mannitol ASS +	1·93	mannitol ASS −	− 0·82
ethanol ASS −	− 0·18	ethanol ASS −	− 0·18
fructose ASS +	1·85	fructose ASS −	− 0·52
Total weight	5·53	Total weight	− 3·59

ASS, ammonium salt sugar.

the figures for the two taxa differed. The weights of evidence in favour of *P. acidovorans* are shown for both positive and negative results in each test and the "swing" which is the sum of the two weights for each test irrespective of their signs. The weights of the test results typical of *P. acidovorans* add up to 5·53 and, as a weight of 3 is equivalent to an identification score of 0·999, strains giving such results should identify. Strains of *P. acidovorans* are found which do not attack fructose ammonium salt sugar (ASS), and as this result reduces the total weight by the swing for fructose (2·37), a weight of 5·53 − 2·37 = 3·16 is obtained so that these strains should still identify. Strains which do not attack mannitol ASS however, will not identify since the total weight for these strains is only 2·78 (5·53 – 2·75 the swing for mannitol) which represents an identification score for *P. acidovorans* of only 0·9983.

For a typical strain of *P. testosteroni* the total weight is – 3·59. Weights less than – 3 indicate that the strain will identify as *P. testosteroni*, and since positive results have not in practice been found for strains of *P. testosteroni* in any of the tests of Table III all the strains of this taxon which we have tested will identify.

CONCLUSION

A computer program incorporating the methods described in this paper is in routine use by our laboratory which provides an identification service for Gram-negative, rod-shaped bacteria. Other programs produce periodic summaries of the data collected during the operation of the service. These summaries are used to review the matrix, the basic data of the identification program, and assess the performance of the program. The test results of the strains examined are available in a computer-readable form for numerical classification studies (e.g. Bascomb *et al.*, 1971). The thorough examination of a strain of bacteria involves a considerable amount of skilled laboratory work and we believe that provided testing methods are standardized, the use of computers to analyse, store and retrieve bacteriological data will become increasingly justified.

ACKNOWLEDGMENTS

We wish to thank the Public Health Laboratory Service for help and support and the Department of Health and Social Security for the grant which made this work possible.

REFERENCES

ANDERSON, J. A. and BOYLE, J. A. (1968). Computer diagnosis: statistical aspects. *Br. med. Bull.* **24**, 230–235.

BASCOMB, S., LAPAGE, S. P., WILLCOX, W. R. and CURTIS, M. A. (1971). Numerical classification of the tribe Klebsielleae. *J. gen. Microbiol.* **66**, 279–295.

BASCOMB, S., LAPAGE, S. P., CURTIS, M. A. and WILLCOX, W. R. (1973). Identification of bacteria by computer: identification of reference strains. *J. gen. Microbiol.* **77**, 291–315.

DICKEY, J. M. (1968). Estimation of disease probabilities conditional on symptom variables. *Math. Biosci.* **3**, 249–265.

GOOD, I. J. (1950). "Probability and the Weighing of Evidence". Griffin, London.

GYLLENBERG, H. (1963). A general method for deriving determination schemes for random collections of microbial isolates. *Ann. Acad. Sci. fenn.* A, IV, **69**, 1–23.

LAPAGE, S. P., BASCOMB, S., WILLCOX, W. R. and CURTIS, M. A. (1973). Identification of bacteria by computer: general aspects and perspectives. *J. gen. Microbiol.* **77**, 273–290.

LEDLEY, R. S. and LUSTED, L. B. (1959). Reasoning foundations of medical diagnosis. *Science* **130**, 9–21.

RYPKA, E. W., CLAPPER, W. E., BOWEN, I. G. and BABB, R. (1967). A model for the identification of bacteria. *J. gen. Microbiol.* **46,** 407–424.

SNEATH, P. H. A. and JOHNSON, R. (1972). The influence on numerical taxonomic similarities of errors in microbiological tests. *J. gen. Microbiol.* **72,** 377–392.

SNELL, J. J. S. and LAPAGE, S. P. (1973). Carbon source utilization tests as an aid to the classification of non-fermenting Gram-negative bacteria. *J. gen. Microbiol.* **74,** 9–20.

WILLCOX, W. R., LAPAGE, S. P., BASCOMB, S. and CURTIS, M. A. (1973). Identification of bacteria by computer: theory and programming. *J. gen. Microbiol.* **77,** 317–330.

9 | New Approaches to Automatic Identification of Micro-organisms

H. G. GYLLENBERG and T. K. NIEMELÄ

Department of Microbiology, University of Helsinki, Finland

Abstract: The principles of automatic identification of micro-organisms are described. In addition to definite identification, consideration of further "identification states" is postulated. The computer-aided procedure can be based on three different mathematical methods, viz. identification by probabilities, identification by Euclidean distances and identification by correlation coefficients. These three methods are described and discussed in detail. The problems of feedback and readjustment in automatic identification systems is also discussed, and finally an experimental illustration of readjustment is given.

Key Words and Phrases: identification of micro-organisms, identification matrix, identification states, reference system, identification limit, neighbourhood limit, identification by probabilities, identification by Euclidean distances, identification by correlation coefficients, readjustment of reference system

INTRODUCTION

Routine identification work constitutes the main activity in many microbiological laboratories. Although certain mechanical devices, such as multipoint inoculators, have been introduced in recent years to increase capacity, the identification procedure is still mainly dependent on manual operation. In cases where the capacity of a laboratory needs to be suddenly increased, e.g. clinical and public health laboratories during outbreaks of food poisonings and/or epidemics, efforts may be restricted by lack of personnel and space. This explains the increase in interest for automatic identification procedures in recent years.

An automatic identification procedure depends on two elements, namely: (1) an automated or semiautomated device or equipment which can handle a large amount of isolate material and read the results of different determinative tests; and (2) computer routines, which process the primary information from

Systematics Association Special Volume No. 7, "Biological Identification with Computers", edited by R. J. Pankhurst, 1975, pp. 121–136. Academic Press, London and New York.

the readings to relevant identification data. The present report deals only with the second of these components of automatic identification of micro-organisms.

BACKGROUND

Identification implies that the unknown object or item to be identified is compared with known objects or items. Accordingly, identification is only possible against a reference system (cf. Rypka and Babb, 1970). As identifications are carried out new information accumulates, and this information can be continuously utilized for a readjustment of the reference system. The whole procedure can then be described as a continuous loop:

(1) establishment of a reference system;
(2) identification against the reference system;
(3) accumulation of new "identification information";
(4) readjustment of the reference system;
(5) identification against the readjusted reference system . . . etc.

The establishment of the reference system involves some method of classification or grouping of previously collected objects or items (since micro-organisms are being dealt with, the individual items can be referred to as strains). There is no restriction on the method of grouping, but since the identification logic is numerical, a grouping based on the principles of numerical taxonomy seems preferable. When the groups in the reference material have been recognized, the frequencies of alternative outcomes are defined for each character of each group. With dichotomous characters (i.e. of the type + or −), the positive alternative can be scored 1, and the negative 0. For continuously varying characters, the readings have to be normalized to correspond to a range of 0–1. With multistate characters, again each state has to be considered separately.

The frequency figures are collected into a group × character-matrix. This matrix constitutes the reference system, and can be referred to as *identification matrix*. An example of an identification matrix for clinical isolates is presented in Table I.

FUNDAMENTAL IDENTIFICATION CRITERIA

The fundamental identification criterion of Dybowski and Franklin (1968) is that a given strain or isolate can be identified as a member of some given group if the probability of belonging calculated for the group in question exceeds a predefined limit. Lapage *et al.* (1970), on the other hand, presume that identification can be accepted if the group in question provides an exclusive alternative. The criterion is that the probability calculated for the group

TABLE I. An example of identification by probabilities: parameters L_{1i}, L_{2i} and b_i ($i=1, \ldots, 6$) and identification matrix for 6 groups and 24 characters

	Group	Limit 1	Limit 2	Best possible log. probability
	1	3·8274	5·2192	1·5016
	2	4·1111	5·5029	1·7852
	3	3·2606	4·6524	0·9348
	4	3·6830	5·0748	1·3571
	5	3·3353	4·7271	1·0095
	6	3·3719	4·7637	1·0461

GROUP CENTRES:

Test	1	2	3	4	5	6
1	0·04	0·00	1·00	1·00	1·00	0·90
2	0·00	0·00	0·90	0·82	0·42	0·90
3	0·00	0·00	1·00	1·00	1·00	1·00
4	0·04	0·24	1·00	0·27	0·17	0·70
5	0·00	0·76	0·00	0·00	0·00	0·00
6	0·74	0·98	0·00	0·00	0·92	0·85
7	0·00	0·59	0·63	0·36	0·92	0·10
8	0·87	0·18	0·00	0·36	1·00	1·00
9	1·00	1·00	1·00	1·00	1·00	1·00
10	0·00	0·00	0·96	0·64	0·75	0·55
11	0·93	0·00	0·00	0·00	0·00	0·00
12	0·04	0·59	0·96	0·00	0·17	0·70
13	0·39	0·35	0·67	1·00	1·00	0·80
14	0·04	0·64	0·96	0·63	1·00	0·85
15	0·09	0·00	0·00	0·00	0·17	0·00
16	0·22	0·00	0·74	0·91	1·00	0·95
17	0·13	1·00	0·89	0·91	1·00	1·00
18	0·57	0·65	0·78	0·91	0·92	0·85
19	0·78	0·94	0·96	1·00	1·00	1·00
20	0·35	1·00	0·96	1·00	1·00	1·00
21	0·04	0·59	0·85	0·18	0·42	0·00
22	0·91	0·94	1·00	1·00	1·00	1·00
23	0·96	0·00	1·00	0·91	1·00	1·00
24	0·83	0·86	0·89	0·91	1·00	0·00
Size:	23	17	27	11	12	20

considered exceeds a predefined relative share of the sum of absolute probabilities for all groups in the identification matrix. This relative share is called normalized probability.

We have combined the identification criteria of Dybowski and Franklin and of Lapage *et al*. Identification as a member of a given group then requires that the unknown strain fulfils two prerequisites, namely:

(1) the *absolute affinity* of the strain to the given group exceeds a predefined limit; and

(2) the *relative affinity* of the strain to the same group must also exceed a predefined limit.

In other words the strain must show high similarity to the group in question, but the strain must not be highly similar to any other group.

"IDENTIFICATION STATES"

The application of the two fundamental identification criteria described above gives rise to some further "identification states" besides definite identification:

(a) Both fundamental identification prerequisites are fulfilled; the strain is *identified*.

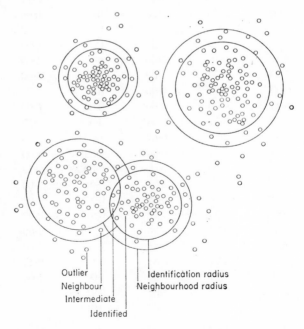

FIG. 1. Geometrical illustration of the four identification states.

(b) The prerequisite of high absolute affinity is fulfilled, but that of exclusivity (i.e. high relative affinity) is not. Accordingly, the strain may fall within two or more groups, and can be referred to as an *intermediate*.

(c) The absolute affinity does not exceed the predefined limit, but is very close to it (and at the same time exclusivity is obvious); the strain can be considered as a *neighbour* to the given group.

(d) Absolute, and often relative, affinities are low to all groups; the strain falls definitely outside the reference system, and can be referred to as an *outlier*.

These four "identification states" are further illustrated in Figs 1 and 2.

ALTERNATIVE IDENTIFICATION PROCEDURES

The following alternative identification procedures have been studied in detail:
(1) identification by the principle of probabilities;
(2) identification based by Euclidean distances; and
(3) identification by correlation coefficients.

Although these procedures have been described elsewhere by the present authors (Gyllenberg and Niemelä, 1974), the main points will be repeated here.

1. Identification by Probabilities

This method presumes that characters considered are qualitative, i.e. that the outcomes of a character constitute mutually exclusive alternatives. This identification principle can be easily programmed for general treatment of dichotomous information, but with multistate characters the programmes need to be "tailor-made" for each purpose although the general principle is the same.

The groups in the reference system are polythetic (also in the sense that all characters are given equal weight in identification). Criteria are thus required to establish how similar the "unknown" to be identified must be, as compared with the mean group characters, in order to be identified as a group member (or to be considered as neighbour to that group). The criteria are given by two probability limits, L_{1i} and L_{2i}, defined for each group in the reference system ($i=1 \dots n$).

These limits correspond to the decision situations: (a) identified or neighbour, and (b) neighbour or outlier. For definition of these limits the intragroup variation has to be considered. The intragroup variation can be described by the maximum probability of belonging to the group. The coefficients c_{ij} which indicate the frequencies of positive character outcomes in the identification matrix at the same time represent the probabilities of the outcomes in question (the probabilities of the corresponding negative outcomes are given by $1 - c_{ij}$). The maximum probability of belonging to a given group is then simply

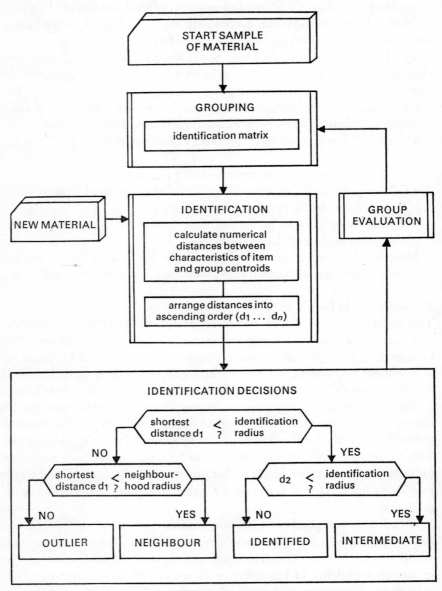

FIG. 2. Schematic illustration of the computer program for identification decisions.

obtained by multiplication of the probabilities of the most frequent alternative outcomes for all considered characters:

$$b_i = \prod_{j=1}^{m} \max \left(c_{ij}, 1 - c_{ij} \right), \tag{1}$$

where m is the number of characters.

A high intragroup variation gives a low value of b, whereas bs for homogeneous groups are high. The limits L_{1i} and L_{2i} are defined by $b_i > L_{1i} > L_{2i}$. Therefore the parameters l_1 and l_2 have been introduced ($l_1 > l_2$), and L_{1i} and L_{2i} are then given by

$$L_{1i} = l_1{}^m \cdot b_i, \text{ and} \tag{2}$$

$$L_{2i} = l_2{}^m \cdot b_i \tag{3}$$

The parameters l_1 and l_2 (which have to be chosen subjectively) show which absolute probabilities of belonging (in relation to the maximum one, b_i) are required as a basis for the conclusions identified or neighbour, respectively. Accordingly, the automated identification decision also requires computation of the absolute probability of the unknown belonging to each of the groups in the reference system (identification matrix):

$$p_i = \prod_{j=1}^{m} f_{ij}, \tag{4}$$

where f_{ij} is the probability of character outcome j in group i. In practice p_i is equal to the total combined probability of the character outcomes for the unknown. For dichotomous characters:

$$= f_{ij} \begin{cases} c_{ij}, \text{ when the unknown behaves positively } (+) \text{ for character } j \\ 1 - c_{ij}, \text{ when the unknown behaves negatively } (-) \text{ for character } j \end{cases}$$

The identification decision, however, also requires evaluation of how much higher p_i is as compared with probabilities found for alternative groups. For this purpose the normalized probability needs to be computed:

$$P_i{}^* = p_i \bigg/ \sum_{j=1}^{n} p_j \tag{5}$$

$P_i{}^*$ indicates the share of group i of the total sum of probabilities for all groups in question (p_j; $j = 1 \dots n$). The decision that an unknown is identified as a member of group i requires that the unknown's absolute probability of belonging to group i is sufficiently high, but also that the normalized probability is high enough to indicate that group i is an exclusive alternative.

If the limit for the normalized probability is set at 0·99 (cf. Lapage *et al.*, 1970), the final identification criteria are

(a) $P_i^* \geqslant 0.99$

(b) $p_i \geqslant L_{1i}$

When condition (b) is fulfilled, but condition (a) is not, then the unknown in question is considered as an intermediate. The normalized probability is not considered for decisions when $p_i < L_{1i}$. The conditions which are valid for the four different "identification states" can be summarized as follows:

Identified	Intermediate	Neighbour	Outlier
$P_i^* \geqslant 0.99$	$P_i^* < 0.99$		
		$L_{2i} \leqslant p_i < L_{1i}$	$p_i < L_{2i}$
$p_i \geqslant L_{1i}$	$p_i \geqslant L_{1i}$		

In practice the absolute probabilities for the different groups in the reference system show a wide range of variance, e.g. from 10^{-1} to 10^{-20}, and, therefore, it is most convenient to express the probabilities as the absolute values of their logarithms, for instance 1–20. These logarithmic figures can also be interpreted as distances, a small figure corresponding to a short distance from the group centroid, and vice versa. Our computer program prints out the probability limits L_{1i} and L_{2i} in a corresponding form, as well as the highest possible probabilities b_i (Table I).

For the unknowns the following information is printed out: label of the strain, the normalized probability for the best group alternatives and the corresponding absolute probabilities (as logarithms). For strains which fulfil the condition $p_i \geqslant L_{1i}$ (i.e. identified and intermediates) the program also lists those characters for which the strain in question differs significantly from the behaviour of the group. Table II illustrates a typical print-out.

2. *Identification by Euclidean distances*

Application of this method implies that identification is based on Euclidean distances between the unknowns and the group centroids. The unknown strains and the group centroids can be considered as points in an *m*-dimensional space (where *m* is the number of characters), and it follows then that the characters should be quantitative (in special cases dichotomous qualitative characters can be accepted). When identification by Euclidean distances is applied the identification matrix is considered to contain the means of the quantitative characters group by group. In order to get an equal weighting of all characters

TABLE II. An example of computer print-out of identification data based on probabilities

Strain	Type	Normalized probability	Best groups/Logarithmic probabilities				Unexpected results
1	Identified	1·0000	1/1·502	2/11·68	6/16·39	4/16·71	
2	Identified	1·0000	1/3·301	2/14·89	6/18·42	4/18·71	
3	Outlier	0·9999	5/5·038	3/9·461	4/9·483	6/9·634	5
4	Neighbour	1·0000	6/4·135	4/10·07	1/10·63	5/11·36	
5	Identified	1·0000	1/2·178	2/12·55	6/14·39	4/17·71	
6	Outlier	0·8667	1/11·00	2/11·84	4/13·13	5/15·09	9 12 21
7	Identified	1·0000	1/1·502	2/11·68	6/16·39	4/16·71	
8	Identified	0·9998	3/0·935	4/4·675	5/5·737	6/8·741	
9	Neighbour	0·9997	6/3·410	5/6·887	4/8·587	3/10·23	20
10	Identified	1·0000	1/2·292	2/12·81	6/15·67	4/18·71	
11	Identified	1·0000	2/2·560	1/11·66	5/12·31	4/13·13	
12	Neighbour	0·9935	5/4·695	3/6·922	6/7·987	4/8·909	5
13	Neighbour	1·0000	1/5·180	2/15·33	4/19·71	6/21·44	12
14	Identified	1·0000	6/1·133	5/6·066	4/7·260	3/8·234	
15	Neighbour	0·8247	5/3·628	4/4·325	6/5·572	1/11·35	14
16	Identified	1·0000	1/1·502	2/11·68	6/16·39	4/16·71	
17	Identified	1·0000	2/2·312	1/12·23	4/18·37	3/18·45	17
18	Outlier	1·0000	6/5·763	1/11·15	3/11·31	4/11·50	
19	Identified	0·9998	3/0·935	4/4·675	5/5·737	6/8·741	
20	Identified	0·9998	3/0·935	4/4·675	5/5·737	6/8·741	
21	Identified	0·9939	6/3·209	4/5·492	5/6·264	3/9·649	
22	Neighbour	0·9998	1/4·023	2/7·949	5/8·515	6/8·522	
23	Identified	0·9998	3/0·935	4/4·675	5/5·737	6/8·741	
24	Outlier	0·9999	3/6·217	4/10·26	6/10·66	1/16·53	14
25	Outlier	0·9994	2/6·673	1/9·918	6/16·26	4/18·39	17

their outcomes have to be modified to fall within the same numerical range
[0, 1]. For dichotomous characters 0 corresponds to a negative outcome ("–"),
and 1 to a positive outcome ("+"). The group centroids are based on the
relative frequencies of positive outcomes.

The intragroup deviation, which has to be considered for identification
decisions, is σ_i. For dichotomous characters the deviations can be calculated
directly from the identification matrix by application of the formula for the
variance of binomial distribution, but in other cases the deviations have to be
calculated in connection with grouping and must be defined for the identification
program in question. For each group two radii can be defined, viz. $l_1 \times \sigma_i$ and
$l_2 \times \sigma_i$ respectively. The parameters l_1 and l_2 correspond to those referred to in
the previous chapter, and, accordingly, the smaller radius ($l_1 \times \sigma_i$) provides the
basis for identification/neighbour decisions, whereas the outer radius determines
neighbour/outlier decisions.

It follows from the definition of the identification radius that if the distance
between two group centroids, i and j, is

$$d_{ij} < l_1 \times (\sigma_i + \sigma_j),$$

groups i and j are partly overlapping. An unknown strain which lies within the
intersection of the both groups thus constitutes an intermediate.

In order to facilitate comparison of different alternatives, the distances d_i
between the unknown strain and the different group centroids can be divided
by the group deviation. This gives normalized distances

$$D_i = d_i/\sigma_i. \tag{6}$$

Denoting the shortest normalized distance D_1 and the next shortest D_2, the
identification conditions can be described as follows:

Identified	Intermediate	Neighbour	Outlier
(a) $D_1 \leqslant l_1$			
	$D_2 \leqslant l_1$	$l_1 < D_1 \leqslant l_2$	$D_1 > l_2$
(b) $D_2 > l_1$			

As in the case of identification by probabilities the parameters l_1 and l_2 are chosen
subjectively, considering, however, that $l_1 < l_2$.

3. Identification by correlation coefficients
This method applies primarily to quantitative characters, but dichotomous
characters can also be used. The similarity between the unknown to be identified

and a group is evaluated by the coefficient of correlation between the characters of the unknown and those of the group. The significance of the correlation coefficient is evaluated by means of the *t*-test. The superiority of the best group alternative can be examined by statistical testing of significant similarity of the two highest correlation coefficients at the selected level of confidence. Our computer program gives the three best group alternatives and the corresponding correlation coefficients as well as the confidence levels in %.

The normalization of the variables (i.e. the characters) can, when desired, be performed by denoting the variable mean$=0$, and the variance$=1$.

In a dichotomous material the variance estimate needed for normalization can be obtained directly from the identification matrix by application of the formula for calculation of the binomial distribution. In other cases the variances must be calculated already when grouping is performed, and the obtained values have to be given to the computer program for identification. The normalization, if performed, has to be carried out both for the identification matrix and for the unknown strains to be identified.

For identification the correlation coefficients between the strain in question and the alternative groups are calculated: r_i $(i=1 \ldots n)$.

$$r_i = \frac{\sum\limits_{j=1}^{m} x_j \cdot c_{ij} - \left(\sum\limits_{j=1}^{m} c_{ij}/m\right) \cdot \sum\limits_{j=1}^{m} x_j}{\left\{\left[\sum\limits_{j=1}^{m} x_j^2 - \left(\sum\limits_{j=1}^{m} x_j\right)^2/m\right] \cdot \left[\sum\limits_{j=1}^{m} c_{ij}^2 - \left(\sum\limits_{j=1}^{m} c_{ij}\right)^2/m\right]\right\}^{1/2}} \tag{7}$$

where m is the number of characters, c_{ij} is the average outcome of character j in group i and x_j is the outcome of character j for unknown strain x.

From r_i the figures t_i and z_i can be calculated:

$$t_i = \frac{r_i\sqrt{m-2}}{\sqrt{1-r_i^2}}, \text{ and} \tag{8}$$

$$z_i = \tfrac{1}{2}\ln\frac{1+r_i}{1-r_i} \tag{9}$$

The two highest correlation coefficients r_I and r_{II} are then considered and from tables included in the program, the corresponding significance figures are defined. On this basis the 0-hypotheses, viz.

  (a) $H_0: r_I=0$ (*t*-test), and

  (b) $H_0: r_I=r_{II}$ (double-sided test)

are either accepted or rejected.

For evaluation of the first hypothesis two limits, T_1 and T_2 are predefined subjectively. These limits correspond to the identification decisions identified/neighbour, and neighbour/outlier, respectively. For the second hypothesis the limit Z is predefined. This limit permits evaluation of the difference between two correlation coefficients. This difference is given by

$$Z = z_1 - z \left/ \sqrt{\frac{2}{m-3}} \right. \tag{10}$$

The conditions for identification decisions are then the following:

Identified	Intermediate	Neighbour	Outlier
(1) $t_I \geqslant T_1$	(1) $t_I \geqslant T_1$		
		$T_2 \leqslant t_I < T_1$	$t_I < T_2$
(2) $z \geqslant Z$	(2) $z < Z$		

This kind of use of correlation coefficient does not satisfy all the statistical assumptions of normal distribution and independence of characters. The correlation coefficient is further somewhat sensitive to changes in coding of character states. However, in spite of weak statistical justification, this coefficient can be used as a technical means of calculating affinities between strains and group centroids.

FEEDBACK AND READJUSTMENT

As has been emphasized in earlier paragraphs, information from identification should be continuously utilized for a readjustment and correction of the reference system, and thus also the identification matrix. Rearrangements may concern both fusion and fission of pre-existing groups, as well as the creation of completely new groups. Parameters for an automated steering of the readjustment process are not easy to define, and the following discussion includes general comments, but no suggestions for an applicable procedure.

Our present approach has been limited to the application of the Euclidean distances, and the characters thus have to be quantitative and/or dichotomous. It seems to us, however, that character properties (i.e. quantitative *vs* qualitative) do not constitute a particular problem. The fundamental difficulty is how various decision-requiring situations in the readjustment process can be defined in terms which allow the formulation of an automated strategy.

In this connection the question arises whether it would be beneficial to try to develop a fully automated readjustment procedure. The alternative is to carry out at least part of the treatment as an interactive man–machine performance, where human control and decision-making is essentially involved. This might concern particularly the creation of new groups as well as the combining of already existing groups, or their splitting into new, smaller groups. Full automation, on the other hand, can be applied to those stages of the procedure where existing groups are adjusted without a change in their number. It would be beneficial for the man–machine work if the whole readjustment procedure could be divided into subsequent but independent elements, each of which could be defined in general terms for the machine. This would facilitate the development of a strongly problem-oriented programming language, and a flexible combination of subprograms to perform the necessary readjustment functions.

At the present stage of our work we use one large program entity which is composed of several shorter modules, and the input element for the control of the program and its modules. The system works by the principle that each new stage of the program uses the results produced in the previous stage or a new information input provided by cards. To start a particular stage in the program the code of this stage and necessary parameters have to be given to the machine. To a system of this kind new modules can easily be added and old modules can be modified without danger of the whole system breaking down. In its present form our system contains the following stages or modules:

(a) input of basic information, and input of information to start later stages;

(b) search for new groups among strains that have been left unidentified;

(c) iterative adjustment of the groups (the number of groups does not change);

(d) identification either by a stable system or cumulation of identified strains into the identification matrix;

(e) calculation of distance distributions for overlapping groups;

(f) print-out of group centroids;

(g) print-out of the distances between group centroids;

(h) print-out of localization of the individual strains, either group by group, or as one list (the latter alternative can be used for print-out on punched cards to provide the input for the next run);

(i) input of the grouping list from the previous print-out;

(j) calculation of the identification matrix and group variances on the basis of the grouping list.

It is worth mentioning that our system in its present form requires that the investigator starts each new stage. The search for new groups, for instance, cannot be performed besides identification, but only at certain intervals decided by the investigator.

An Illustration of the Re-adjustment of the Reference System

The basic material was composed of 150 strains representing clinical isolates. The grouping was based on conventional taxonomy, i.e. the strains were grouped according to the names assigned to them in the clinical laboratory where isolation was carried out. The strains were distributed to groups as follows:

Group	Label	Number of strains
Pseudomonas	1	25
Proteus	2	21
E. coli	3	38
Klebsiella	4	31
Enterobacter	5	20
Acinetobacter	6	5
Alcaligenes	7	3
Providencia	8	4
Citrobacter	9	3

This material was utilized to define a reference system, and intragroup variances were estimated. Groups 1 and 9 were the most homogeneous, whereas groups 6 and 7 were the most heterogeneous. For the evaluation of changes in the system the identification matrix and intergroup distances were printed out. Groups 6 and 7, and 5 and 9, respectively, were closest to each other.

With the reference system available more material, altogether 333 strains, including the 150 above-mentioned strains, was identified. The results of the identification are condensed into Table III.

As can be seen from the table the number of intermediates is extremely high for groups 1 and 7. This depends on small intergroup distances and the high

TABLE III. Identification of 333 clinical isolates in a reference system based on non-phenetic conventional classification

Group	Identified	Intermediate	Neighbour	Outlier	Σ
1	3	46			49
2	35		11		46
3	61	2	16	2	81
4	44	8	7		59
5	39	5	5		49
6	9	7	2		18
7	10	10	1		21
8	4		2		6
9	1	3			4
Σ	206	81	44	2	333

variance in group 7 which produces a large identification radius for this group.

As a further treatment step all strains located inside the identification radii of the different groups were considered and the group centroids and intragroup variances were recalculated. Among those strains which fell outside the identification radii (i.e. neighbours and outliers; 46 strains in total) new groups were searched for. 4 new groups were recognized:

Group	
10	12 strains—*E. coli*
11	5 strains—*Proteus*
12	5 strains—*Klebsiella*
13	{ 2 strains—*Enterobacter* 1 strain —*E. coli*

The new groups were far enough from each other and from the "old" groups. Since new groups had been found it was considered necessary to carry out an iterative adjustment of the grouping—(as mentioned above the procedure started from a non-phenetic traditional grouping). After three iterations the grouping was stabilized to the following distribution of the strains among the 13 groups now present.

Group	Description	Number of strains	Remarks
1	*Pseudomonas*	59	2 strains labelled *Alcaligenes* were included
2	*Proteus*	32	
3	*E. coli*	63	
4	*Klebsiella*	47	1 *E. coli* included
5	*Enterobacter*	44	2 *E. coli* included
6	*Acinetobacter*	12	
7	*Alcaligenes*	13	{ 2 *Pseudomonas* and 3 *Acinetobacter* included
8	*Providencia*	6	
9	*Citrobacter*	6	
10	*E. coli*	15	The "new" *E. coli* group
11	*Proteus*	13	The "new" *Proteus* group
12	*Klebsiella*	10	The "new" *Klebsiella* group
13	Miscellaneous	6	{ 3 *Enterobacter* 2 *Klebsiella* 1 *E. coli*

In the old groups the variance was reduced except for Groups 8 and 9 to which a few new strains were added. The largest variance was found for Group 11, but even for this group the variance was smaller than the largest variances for the original groups. Accordingly the treatment for adjustment of the groups resulted in more homogeneous groups.

Finally a new identification matrix was produced, and the whole material was re-identified in order to evaluate the effect of regrouping. At the same time the identification parameters l_1 and l_2 were somewhat modified (decreased); 83·5% of the strains now fell within the identification radii, showing the following distribution:

Group	Description	Number of strain	Remarks
1	*Pseudomonas*	48	
2	*Proteus*	24	
3	*E. coli*	50	
4	*Klebsiella*	40	
5	*Enterobacter*	40	4 *E. coli*
6	*Acinetobacter*	9	
7	*Alcaligenes*	16	5 *Pseudomonas*, 3 *Klebsiella*, 1 *Citrobacter*
8	*Providencia*	6	
9	*Citrobacter*	5	
10	*E. coli*	12	
11	*Proteus*	13	
12	*Klebsiella*	9	
13	Miscellaneous	6	3 *Enterobacter*, 2 *Klebsiella*, 1 *E. coli*

REFERENCES

DYBOWSKI, W. and FRANKLIN, D. A. (1968). Conditional probability and the identification of bacteria. A pilot study. *J. gen. Microbiol.* **54**, 215–229.

GYLLENBERG, G. H. and NIEMELÄ, T. K. (1974). Basic principles in computer-assisted identification of micro-organisms. *In* "New approaches to the identification of micro-organisms", Chapter 13, Wiley, New York.

LAPAGE, S. P., BASCOMB, S., WILLCOX, W. R. and CURTIS, M. A. (1970). Computer identification of bacteria. *In* "Automation, Mechanization and Data Handling in Microbiology" (A. Baillie and J. R. Gilbert, eds), pp. 1–22. Academic Press, London and New York.

RYPKA, E. W. and BABB, R. (1970). Automatic construction and use of an identification scheme. *Med. Res. Engng.* **9**, 9–19.

10 | Simulation of a Computer-aided Self-correcting Classification Method

T. K. NIEMELÄ and H. G. GYLLENBERG

Department of Microbiology, University of Helsinki, Finland

Abstract: A simulation of a computer-aided continuously self-correcting classification process is described. The authors' conclusion is that a formalization of a numerical process of this kind provides a better tool for development of a useful mocrobial taxonomy than random *de novo* classifications.

Key Words and Phrases: identification of micro-organisms, reference systems for identification, identification states, self-correction of classifications, de novo-classifications, simulation of classification and classification

INTRODUCTION

The purpose of this paper is to illustrate by graphical means the technique of a classification which continuously corrects itself, and hence also improves the basis for identification. In the present case, imaginary material is exploited, but we want to emphasize that the same technique has been applied to many actual experimental materials (e.g. Gyllenberg *et al.*, 1974). The technique in question allows treatment of materials which are: (a) *continuous in time*, which implies that the whole material cannot be treated in one batch, and that obtained new information is applied to adjust earlier conclusions on classification and identification; and (b) *quantitatively* so *large* that hierarchical classification methods are out of question as requiring too much computer time and memory.

A special feature of our method is that it requires continuous observation and decisions from the side of the investigator. Accordingly, it is supposed that the computer program provides numerical information of different kinds. Obviously our method would be most successful in an interactive man–machine procedure, but for the moment the computer devices at our disposal have not made this approach possible.

Systematics Association Special Volume No. 7, "Biological Identification with Computers", edited by R. J. Pankhurst, 1975, pp. 137–151. Academic Press, London and New York.

The procedure, in its present design, can be divided into the following main steps:

1. A part of the material (information) is used for an initial classification, e.g. by a hierarchical numerical method, or, alternatively, by a non-numerical method. The outcome of this classification provides an initial reference system.

2. Further material (information) is considered and treated as described under steps 3–6.

3. The new material is accumulated into the "old" groups within the limits of the selected similarity, and using subsequent iterations, until a state of stability is achieved.

4. Among the individuals left outside the "old" groups new groups are searched for. If "new" groups are recognized, further iterative adjustment can be carried out.

5. If considered worthwhile the material, or a part of it, can be "identified" against the reference system at one's disposal. Identification provides detailed numerical information.

6. Lumping and splitting off of individual groups is considered on the basis of calculation of distances between group centroids and of intragroup variances, and by checking both intra- and intergroup distance distributions of individual items.

As new material is introduced the whole procedure is repeated starting from step 3 above.

MATERIALS AND METHODS

The material is imaginary, and for the sake of simplicity, it is described by a two-dimensional model.

A complete computer-simulation has been carried out. Numerical classification by complete linkage was applied to obtain the initial classification, and for identification purposes the automatic identification procedure based on Euclidean distances was chosen. Full details on identification principles and alternative identification methods have been given by Gyllenberg and Niemelä (1974 and this volume), whereas an application of the present approach of self-correcting classification to clinical isolates has been reported by Gyllenberg et al. (1974).

The applied identification logic is based on a *reference system*. The self-correcting classification thus implies a continuous readjustment of this reference system. Identification against the reference system gives rise to four mutually exclusive "identification states", viz. "*identified*" (as a member of one group in the reference system), "*intermediate*" (identification as a member of one or more groups in the reference system), "*neighbour*" (not identified as a member of groups defined by the reference system, but showing close correlation to one group), and "*outlier*" (recognized as not related to any group in the reference system). Reference to identification and/or neighbourhood limits in the following text thus indicates conditions for conclusions relating to the identification states "identified" and "neighbour", respectively.

RESULTS

Initially there were 50 items or taxonomic units defined in a two-dimensional representation as indicated in Fig. 1. The frequency distribution of the distances between all possible pairs of the first 50 taxonomic units in our example are given in Fig. 2. In Fig. 2 the first high peak indicates the occurrence of at least one tight cluster. The following less pronounced peak indicates that the material may contain more than one tight cluster.

Application of complete linkage-analysis to the 50 first taxonomic units indicates that eight clusters can be distinguished in the material (Fig. 3). The identification of these eight clusters in the original distribution figure is illustrated in Fig. 4.

As a further step 50 new taxonomic units were introduced (Fig. 5), and the combined distribution of the total of 100 units is given in Fig. 6. The 50 new units were added to the existing eight groups by identification against these eight groups as a reference system. This gave rise to changes in group centroids as well as in intra- and intergroup variances. An iterative adjustment to place each unit in its closest group was therefore performed. The results are given in Fig. 7. As can be seen from this figure, there now occur several units which fall outside groups 3 and 4, and group 6, respectively.

In fact two new groups were recognized at this stage, and, accordingly, the reference system was amended to include ten groups instead of eight (Fig. 8). The situation indicated in Fig. 8 can be considered as a direct development from the earlier grouping which was based on less information. It is, however, also possible to produce a completely new classification at any stage of information collection. The outcome of a *de novo* classification of the same material is indicated in Fig. 9. The differences between the two exercises (Figs 8 and 9, respectively) are not particularly significant, but the problem definition is quite interesting, and it can obviously be concluded that numerical taxonomy is not necessarily free from a previous "history", and that for the numerical approach also, the choice is between a continuously "correcting" taxonomy in comparison with *de novo* systems.

Figure 10 illustrates the identification- and neighbourhood-limits of the first ten groups based on the first 100 taxonomic units in the material. This

T. K. Niemelä and H. G. Gyllenberg

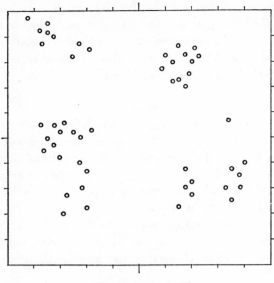

FIG. 1. Units 1–50.

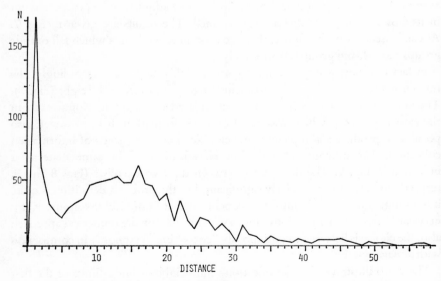

FIG. 2. Frequency distribution of distances between pairs. Units 1–50.

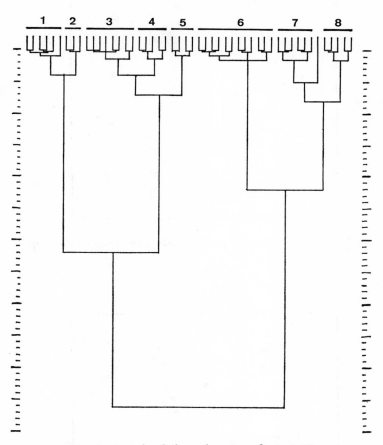

FIG. 3. Complete linkage clustering of units 1–50.

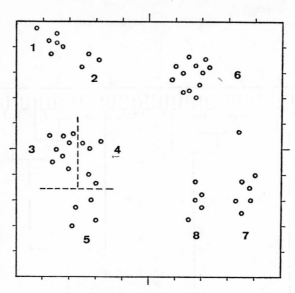

Fig. 4. Units 1–50 divided by complete linkage clustering into 8 groups.

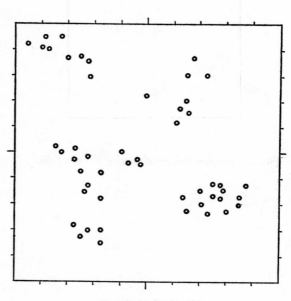

Fig. 5. Units 51–100.

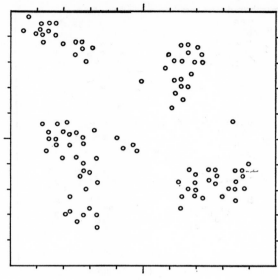

Fig. 6. Units 1–100.

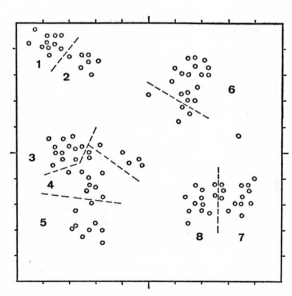

Fig. 7. Units 1–100 accumulated into groups 1–8.

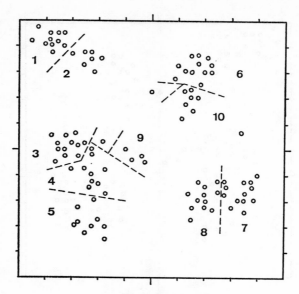

FIG. 8. Units 1–100 in groups 1–10.

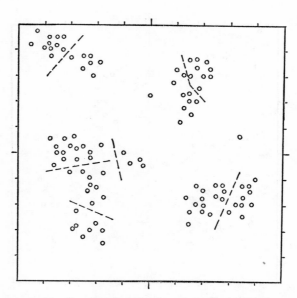

FIG. 9. Complete linkage clustering of units 1–100.

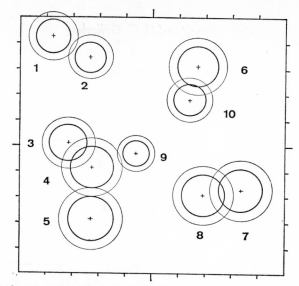

FIG. 10. Identification and neighbour limits of groups 1–10 according to units 1–100.

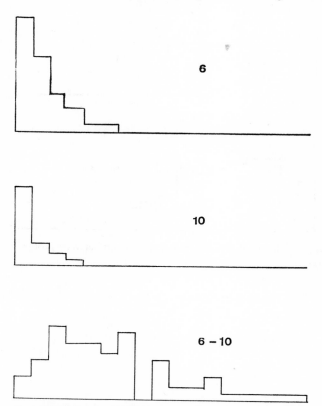

FIG. 11. Frequency distributions of distances within and between groups 6 and 10 (units 1–100).

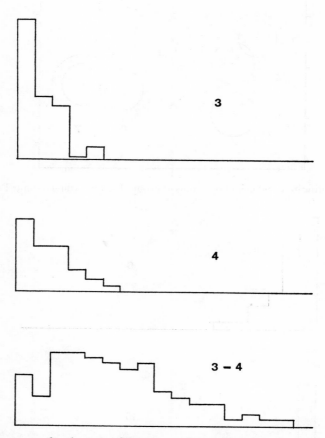

FIG. 12. Frequency distributions of distances within and between groups 3 and 4 (units 1–100).

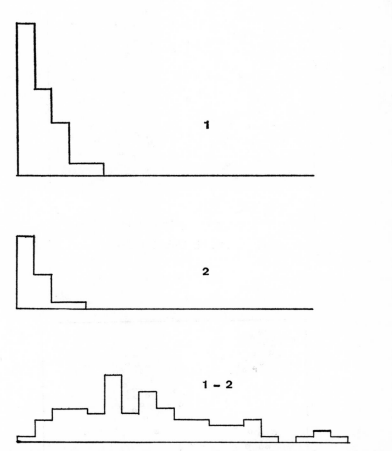

FIG. 13. Frequency distribution of distances within and between groups 1 and 2 (units 1–100).

T. K. Niemelä and H. G. Gyllenberg

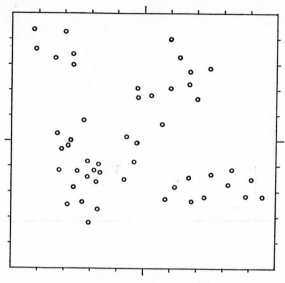

Fig. 14. Units 101–150.

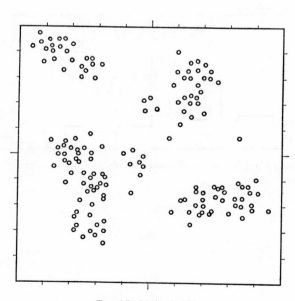

Fig. 15. Units 1–150.

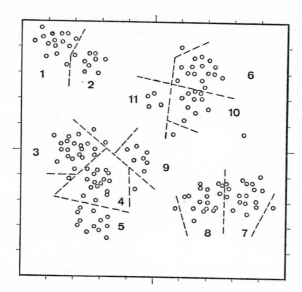

FIG. 16. Units 1–150 divided into 11 groups.

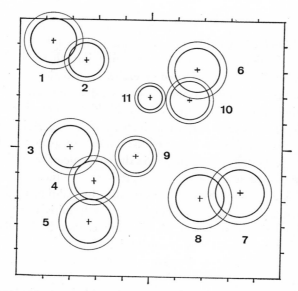

FIG. 17. Final identification and neighbour limits of groups 1–11 according to units 1–150.

representation considers the groups as primary units in comparison with the level of individuals dealt with in the earlier figures. It is obvious from Fig. 10 that certain groups, e.g. 6 and 10, 3 and 4, and 1 and 2, respectively, are situated very close to one another. The problem thus arises whether such close groups should not be lumped together; therefore an analysis has been performed of the frequency distributions of distances between individuals in these groups. Theoretically, a uniform group should give only one peak in the distribution curve, whereas the existence of two distinct groups should be indicated by two peaks, one at a short distance indicating the occurrence of distinct clusters, and one at a longer distance, indicating the occurrence of several strains at a considerable distance from each other.

As can be seen from Fig. 11 the theoretical presumption is quite obvious for groups 6 and 10. For groups 3 and 4 (Fig. 12) the situation is more problematical, whereas a separation into two groups can be assumed (Fig. 13) for groups 1 and 2.

Figure 14 illustrates the distribution of a further group of 50 taxonomic units, whereas Fig. 15 gives the cumulative distribution of all the 150 taxonomic units considered in this study. New cumulation and iterative adjustment created one new group, and the situation following the relevant treatment is illustrated in Fig. 16. The identification and neighbourhood limits with 11 groups in the reference system are given in Fig. 17.

DISCUSSION

The present simulation experiment outlines a procedure for a continuously self-correcting classification. The results obtained may show that a classification (and thus also identification) which is based on earlier information can be formalized and hence steered by computer-aided processes. The fundamental assumption is that a continuous classification process of this kind approaches stability. The exercise reported above indicated that an increase of the underlying experimental material from 50 to 150 items increased the number of groups in the reference system from 8 to 11. The actual change in the coverage of the reference system is indicated in Fig. 18. It can be seen that certain groups (i.e. 1, 2, 3, 6, 7, and 8) were already well defined by the first 50 units. In addition to the three new groups, a considerable change has occurred only in the initial groups 4 and 5.

Since a simulation experiment of this kind cannot be based on random "production" of units, but needs a contribution from the experimenter, we did not consider a continuation of the experiment to be worthwhile. Nevertheless, we assume that a system of this kind approaches stability, and that a continuous self-correcting classification, based on computer-aided numerical processes, may produce an optimal grouping and also describes the earlier development up to the stage when stability is attained.

The alternative here is *de novo* classification. However, this introduces the drawback of lack of, or unintelligible, reference to earlier classifications. A computer-aided, continuously self-correcting classification, on the contrary, can be applied to all kinds of initial reference systems, and, accordingly, it seems to provide a useful link between traditional and numerical approaches to classification and identification of micro-organisms.

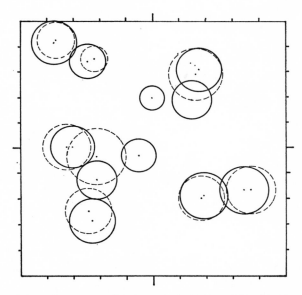

FIG. 18. Identification limits according to units 1–50 (dotted lines) and units 1–150 (solid lines).

REFERENCES

GYLLENBERG, H. G. and NIEMELÄ, T. K. (1974). Basic principles in computer-assisted identification of micro-organisms. *In* "Computer-assisted approaches to the identification of micro-organisms", Chapter 13. John Wiley, New York.

GYLLENBERG, H. G., NIEMELÄ, Y. K., JOUSIMIES, H., HATAKKA, A., SEDERHOLM, H. and NIEMI, J. S. (1974). Development of reference systems for automatic identification of clinical isolates of bacteria. Report No. 9 from the Department of Microbiology, University of Helsinki, Finland. (Also to be published 1975 in *Archives of Immunology and Experimental Therapy* (Poland)).

11 | Pattern Recognition and Microbial Identification

EUGENE W. RYPKA

*Department of Microbiology, Lovelace Center for Health Sciences,
Albuquerque, New Mexico, U.S.A.*

Abstract: The bases for the data sets of microbial identification schemes are considered in regard to traditional, numerical, and primary classifications. Traditional microbial data sets have a low information content and the disjointness of organisms is poor (ranges 46·429–88·653% for 79178 possible pairs of organisms [908] in eight subgroups). The optimal character sets for partitioning the disjoint pairs of organisms has been determined for eight subgroups. Numerical microbial data sets have excellent information content but the sets appear to be small comparatively speaking. A limiting factor of numerical taxonomy for large data sets may be pair comparison of strains over their character states. Primary data sets are illustrated with an on-line method of trapping data continuously and of clustering these data. By horizontal and vertical rearrangement of columns (characters) and rows (organisms) of character state data (in binary variables or number systems to any radix) to fit truth tables, clusters and hierarchial order are obtained. The general identification equation $S = [1 - v^{-k}]/[1 - v^{-c}]$ was developed in the context of truth table clustering.

Key Words and Phrases: pattern recognition, microbial identification, classification, identification, microbial data sets, classification, traditional, classification, numerical, classification, primary, matrix learning, concept learning, truth table classification, truth table identification, continuous classification, identification equation

"Instead of just asking the traditional question of how human minds come to learn from experience, this study asked how one could design a system that would learn from its experience in some 'optimal fashion'."

C. West Churchman
(in *Challenge to Reason*)

INTRODUCTION

Pattern recognition may mean something quite different to the botanist and to the microbiologist—at least in terms of methodology and of subject matter—

Systematics Association Special Volume No. 7, "Biological Identification with Computers", edited by R. J. Pankhurst, 1975, pp. 153–180. Academic Press, London and New York.

from what it means to the psychologist, electronic engineer, computer scientist, physicist, and mathematician. Humans usually are good visual and auditive pattern recognizers. Although indications to the contrary are evident, biologists still depend largely upon their own senses to recognize specimens whereas the physical scientists have developed many elegant optical and auditive information processing systems to recognize their "specimens", for example, fingerprints, speech, faces, printed characters, and clouds (Duda and Hart, 1973; Firschein et al., 1974; Gose, 1969; Grusser and Klinke, 1971; IEEE, 1972; Meisel, 1972). The importance of understanding pattern recognition, in general, for biology and specifically for areas of artificial intelligence involving heuristics and language translation (Firschein et al., 1974; IEEE, 1972) cannot be underestimated. In both contexts, biology and artificial intelligence, pattern recognition, if desired, may be considered to be in the realm of cybernetics especially in regard to the possibility of amplifying machine intelligence (Ashby, 1970).

From the context of the subject matter in the physical sciences it generally is possible to determine the meaning of pattern recognition. Many investigators consider pattern recognition to be synonymous with classification and state that properties of input stimuli or of input signals are used in classification or pattern recognition to obtain groups (Gose, 1969). In an excellent article concerning methodologies of pattern recognition, Watanabe (1969) considered clustering and pattern recognition as "tasks of classification" and recognition of patterns as an inductive process. Some terminological confusion may exist between the biological and physical sciences. Kanal and Chandrasekaran (1969) have commented upon this use of terminology and have made some important distinctions. Recognition of patterns by machine has two different but sometimes interrelated goals. On the one hand are the attempts by mechanical means to simulate human recognition and perception of patterns. On the other hand are automatic processes for handling data which may be used for creating some classification work. They emphasize also that certain human mental activities are programmed and mechanized by methods entirely different from those performed by machines. This point may seem obvious and trivial until one considers the excellent work (Minsky, 1968; Minsky and Papert, 1969) but the inherent difficulties in development, in part by analogical reasoning, of artificial intelligence. At times, "The brain is like a computer", at other times, "The computer is like the brain"—and the similies and metaphors indicate that duplication of human intelligence is extremely difficult. A problem may be that the mechanical devices are attempting to duplicate exactly and to recognize patterns the human brain actually "sees" when the device should be programmed to present perceptually and conceptually different patterns in a uniquely classified form at one time. To this end, perceptual and conceptual clustering methodology has been devised by which the two goals—simulation of patterns and a classificatory work—possibly may be presented as one large pattern and therefore better understood. In this context, pattern recognition of phenotypic characters of microbial specimens for purposes of unique identification will be discussed. To avoid further terminological confusion, pattern recognition will be considered synonymous with identification, determination, and diagnosis, and these terms will be considered inverse mental activities to classification (Rypka, 1971). We must classify before we can identify. Therefore, different classificatory data sets in microbiology from which identification schemes are constructed will be described briefly in the beginning. The reason is simple: unless classifications are disjoint, identification of future isolates and unique elements can never be accomplished.

THE CLASSIFICATORY DATA SETS FOR TRADITIONAL, NUMERICAL, AND PRIMARY IDENTIFICATION SCHEMES

1. Traditional Schemes

Traditionally, in microbiology, data used for constructing determinative or identification schemes are classificatory data. These types of information are obtained from relatively small data sets of the classifications of bacteria which are conjoint closely in terms of phenotypic character states. Hierarchical order is suggested and a larger classification of the bacteria obtained by the piecing and linking together of the smaller data sets. Analysis of determinative schemes such as those in "Bergey's Manual of Determinative Bacteriology" (Breed *et al.*, eds, 1957) indicates that partitioning of the elements in the classification is not adequate and that characteristics are used with ambiguous and vague end-points of tests for detection of characters. The results of this type of analysis of pair comparison of elements by characters over half of the heterotrophic bacteria described in Bergey's Manual are shown in Table I. Identification schemes constructed from classificatory data obtained by so-called traditional methods are designated *traditional* schemes.

2. Numerical Schemes

At the present time there is extensive literature in numerical taxonomy. After the publication of Sneath's work (1957a, b) and the work of Michener and Sokal (1957) and Sokal and Michener (1958) the field has grown rapidly for over a decade. There are adequate textbooks including those by Sokal and Sneath (1963), Cole (1969), Jardin and Sibson (1971) and Blackith and Reyment (1971).

TABLE I. An analysis of 908 heterotrophic bacteria over 724 of their characteristics. The organisms are placed into eight subgroups depending upon their oxygen requirement (anaerobic means strict anaerobe and aerobe means strict aerobe, facultative anaerobe, or microaerophile), Gram staining characteristics, and cell morphology (coccus means spherical cell and bacillus means rod, vibrio, spirillum or filamentous form). Depending upon character variation for one or more of these three characters, an organism could be placed in all eight subgroups. The "Per Cent Pairs Separated" is shown in the middle column. Information content was determined for each subgroup. For example, for subgroup 1, the total information content would be 728 (8×91) and the analysis indicates the per cent of characters stated as being present, absent, variable, or with no information. This analysis is reasonably complete (Rypka, 1971). See Fig. 1 for minimal character sets for each subgroup. As indicated by this analysis, the traditional classificatory data set has a low information content, has many nondisjoint pairs of species, and must have determinative schemes inadequate to identify future isolates

Subgroups[a]	Original Matrix Dimensions[b]	Pairs Separated By Characters Used	Pairs To Separate[c]	Per cent Pairs Separated	For Each Subgroup, Information Content (Per cent)[d]			
					1	0	2	-2
1. Anaerobic Gram negative cocci	8x91	13/	28	46.429	14.16	22.11	4.15	59.59
2. Anaerobic Gram negative bacilli	80x118	2256/	3160	71.424	14.51	17.03	4.82	63.55
3. Anaerobic Gram positive cocci	42x88	506/	861	58.769	12.87	18.67	4.65	63.81
4. Anaerobic Gram positive bacilli	181x166	14433/	16290	88.613	14.29	12.48	4.41	68.83
5. Aerobic Gram negative cocci	82x95	2746/	3321	82.716	2.45	3.24	1.02	93.29
6. Aerobic Gram negative bacilli	244x252	26296/	29646	88.653	1.81	1.55	0.79	95.85
7. Aerobic Gram positive cocci	144x157	8999/	10296	87.354	1.34	2.32	0.74	95.61
8. Aerobic Gram positive bacilli	177x166	12757/	15576	81.947	6.69	6.16	3.28	83.86

[a] Analysis covers 908 heterotrophic organisms and 724 characters. An organism may occur in more than one subgroup.

[b] Matrix dimensions refer to the number of organisms by the number of characters. Only characters which, for at least one organism in the matrix, have a character definitely described as being absent, variable, or present are included.

[c] Pairs to separate = G(G-1)/2 (G= no. of organisms in the subgroup).

[d] 1 = characters present, 0 = absent, 2 = variable, -2 = no information.

Sokal and Sneath (1963) and Sneath and Sokal (1973) include development of identification schemes as part of taxonomy.

In numerical taxonomy, character variables are considered to have equal weight in the classificatory process—that is—on an *a priori* basis some variables are *not* considered to be more important than others and hence are not given more or less weight in terms of determining the conjointness of elements. Every element in the classificatory data set is compared with every other element over the states of the character variables and the number of similarities and the number of differences of states of like characters are determined. From these data many types of coefficients of association, similarity, or conjointness have been devised (Sneath, 1957b; Sokal and Sneath, 1963). After determination of similarity of pairs of elements, some form of synthesis of elements into classes is done. Pair comparison of elements may be a limiting factor of numerical taxonomy when this methodology is applied to large data sets. This comparison, however, serves to indicate if the classified elements are disjoint for purposes of identifying future isolates.

3. Primary Schemes

In 1963, Gyllenberg described a method for group construction from isolates, and after group construction, a method for determining a minimal test set for characters* to be used to determine or to identify future isolates as being members of the groups. Importantly, the necessity was realized for constructing a classification directly from primary data. Use of primary data directly for classification and identification tasks covered an area not included in traditional or new methodological concepts. Gyllenberg (1963, 1964) observed that taxonomic studies usually were concerned with limited numbers of strains of bacteria with a small range of disjointness. In some forms of microbiological research, however, determinative studies are required in which there are thousands of isolates which may have a wide variation in conjointness. Identification of the isolates by traditional methods would be too time-consuming and it was considered desirable to develop a method for finding the smallest number of tests for characters for partitioning the elements. The necessity for sufficient information was emphasized. *Identification schemes* constructed from primary classificatory data are designated *primary schemes.*

IDENTIFICATION SCHEMES CONSTRUCTED FROM TRADITIONAL CLASSIFICATORY DATA

1. Octals

Initially, traditional classificatory data were retrieved from the literature, coded numerically and automatically sorted, summarized, and spread into organism *vs* character matrices. The information content of eight of these submatrices shown in Table I (Rypka, 1972) varies from about 4 to 40%. For purposes of constructing identification schemes, characters were selected sequentially in the following order: the first character separated or partitioned the most pairs of organisms, the second selected character partitioned the most organisms conditionally independent of the first character, and so forth, until an optimal character set

---

* The minimal character set of Gyllenberg actually refers to the theoretical minimal number of characters required to separate or to partition the elements in the matrix. Using binary variables $T_{min} = \log_2 G$ (G = number of elements in the data set). In actual practice, with real world data sets, it is usually necessary to select many more characters to separate the elements. Optimal character set refers to selection of an arbitrary number of characters based upon an investigator's requirements for the use of his identification scheme.

TABLE II. Three examples of truth tables for three different values of logic. Each column of each truth table contains an equal number of each type of symbol. That is, in the two-valued or binary logic table, $2^5 = 32$, each column contains 16 ones and 16 zeros; in the three-valued table, $3^3 = 27$, there are 9 twos, 9 ones, and 9 zeros in each column; in the five-valued table, $5^2 = 25$, each column has 5 fours, 5 threes, 5 twos, 5 ones, and 5 zeros. The value states of the first column of each table divide each table into the number of parts equal to the value of logic, that is, halves ($2^5 = 32$), thirds ($3^3 = 27$), or fifths ($5^2 = 25$). In turn, and successively for each column, further division of each part occurs. This work describes the method of rearranging real data sets to fit most closely truth tables of n-elements and n-characters. In determinative efforts, this provides that identification or diagnosis is proceeding at an optimal rate providing test sensitivity for detecting character states is neither vague nor ambiguous. An importance of many-value logic compared to binary variables is that with binary variables ideally each variable excludes half of the possibilities as identification proceeds whereas with five-value logic, for example, each variable excludes 80% of the possibilities as identification proceeds

TWO-VALUED LOGIC $2^5 = 32$	THREE-VALUED LOGIC $3^3 = 27$	FIVE-VALUED LOGIC $5^2 = 25$
1 1 1 1 1 1	1 2 2 2	1 4 4
2 1 1 1 1 0	2 2 2 1	2 4 3
3 1 1 1 0 1	3 2 2 0	3 4 2
4 1 1 1 0 0	4 2 1 2	4 4 1
5 1 1 0 1 1	5 2 1 1	5 4 0
6 1 1 0 1 0	6 2 1 0	6 3 4
7 1 1 0 0 1	7 2 0 2	7 3 3
8 1 1 0 0 0	8 2 0 1	8 3 2
9 1 0 1 1 1	9 2 0 0	9 3 1
10 1 0 1 1 0	10 1 2 2	10 3 0
11 1 0 1 0 1	11 1 2 1	11 2 4
12 1 0 1 0 0	12 1 2 0	12 2 3
13 1 0 0 1 1	13 1 1 2	13 2 2
14 1 0 0 1 0	14 1 1 1	14 2 1
15 1 0 0 0 1	15 1 1 0	15 2 0
16 1 0 0 0 0	16 1 0 2	16 1 4
17 0 1 1 1 1	17 1 0 1	17 1 3
18 0 1 1 1 0	18 1 0 0	18 1 2
19 0 1 1 0 1	19 0 2 2	19 1 1
20 0 1 1 0 0	20 0 2 1	20 1 0
21 0 1 0 1 1	21 0 2 0	21 0 4
22 0 1 0 1 0	22 0 1 2	22 0 3
23 0 1 0 0 1	23 0 1 1	23 0 2
24 0 1 0 0 0	24 0 1 0	24 0 1
25 0 0 1 1 1	25 0 0 2	25 0 0
26 0 0 1 1 0	26 0 0 1	
27 0 0 1 0 1	27 0 0 0	
28 0 0 1 0 0		
29 0 0 0 1 1		
30 0 0 0 1 0		
31 0 0 0 0 1		
32 0 0 0 0 0		

FIG. 1. Character sets optionally selected for separation of pairs of elements shown in the subgroup matrices of Table I. These schemes are exhaustive and are applicable to identification of over half the described bacteria. These schemes are based upon the traditional classificatory data set and indicate that it is impossible using this data set to obtain accurate identification of future encountered bacteria because of initial lack of disjointness of classification coupled with inadequate sensitivity of tests to detect phenotypic characters in some cases. The acronyms for the subgroups were suggested by Robert Babb. Matrices of real data or organism versus the test sets fit most closely the partitioning truth tables of which smaller examples are shown in Table II. The minimal character set was limited arbitrarily to nine tests and is referred to more precisely as the optimal set.

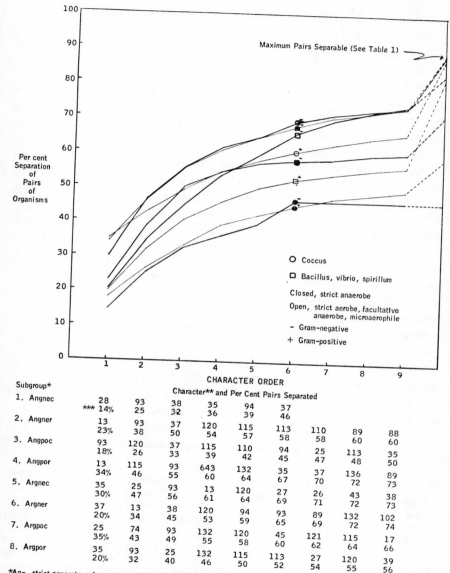

Per cent
Separation
of
Pairs
of
Organisms

Maximum Pairs Separable (See Table 1)

O Coccus

□ Bacillus, vibrio, spirillum

Closed, strict anaerobe

Open, strict aerobe, facultative
anaerobe, microaerophile

− Gram-negative

+ Gram-positive

CHARACTER ORDER

Subgroup*	Character** and Per Cent Pairs Separated								
1. Angnec *** 14%	28	93	38	35	94	37			
		25	32	36	39	46			
2. Angner 23%	13	93	37	120	115	113	110	89	88
		38	50	54	57	58	58	60	60
3. Angpoc 18%	93	120	37	115	110	94	25	113	35
		26	33	39	42	45	47	48	50
4. Angpor 34%	13	115	93	643	132	35	37	136	89
		46	55	60	64	67	70	72	73
5. Argnec 30%	35	25	93	13	120	27	26	43	38
		47	56	61	64	69	71	72	73
6. Argner 20%	37	13	38	120	94	93	89	132	102
		34	45	53	59	65	69	72	74
7. Argpoc 35%	25	74	93	132	120	45	121	115	17
		43	49	55	58	60	62	64	66
8. Argpor 20%	35	93	25	132	115	113	27	120	39
		32	40	46	50	52	54	55	56

*An-, strict anaerobe   Ar-, strict aerobe, facultative anaerobe, microaerophile
-g-, Gram   -ne-, negative · -po-, positive   -c, coccus   -r, bacillus, vibrio, spirillum, filamentous

**13 Motility, 17 Grow on 40% bile blood agar, 25 Catalase, 26 Oxidase, 27 Urease, 28 Gelatinase, 35 Nitrate
reduction, 37 Indole, 38 H_2S production, 39 Litmus (acid), 43 Citrate, 45 Acetylmethylcarbinol,
74 Arginine dihydrolase, 88 Arabinose, 89 Xylose, 93 Glucose (acid), 94 Glucose (gas), 102 Rhamnose,
110 Fructose, 113 Maltose, 115 Lactose, 120 Sucrose, 121 Raffinose, 132 Mannitol, 136 Glycerol,
643 Spore position.

***Rounded to nearest whole number

was determined; or, if desired, any number of characters could be selected (Rypka and Babb, 1970a, b). In order, the character states of the character set for each organism were traced simultaneously over a truth table (Rypka et al., 1967); the size of the table was based upon the number of tests in optimal character set. Examples of truth tables are shown in Table II. Variable character states are considered to agree with any character state and would allow the organisms to be sorted into as many rows of the truth table wherever the character state data agreed with the row pattern of binary variables. If sub-groups of organisms were formed by the sorting process, the method of selecting another test for character set from the unselected characters would be done and identification would proceed in a stepwise manner. The character sets for the eight subgroups of bacteria in Table I are shown in Fig. 1 (Rypka and Babb, 1970b). Figure 1 indicates also the rate of separation of pairs of organisms by the character sets. The method of construction of determinative schemes creates every possible answer for the selected determinative character set for all organisms in the scheme. With time, and in attempts to extend information (Ashby, 1970), patterns of results occurring temporarily and geographically are learned. From this knowledge the schemes may be reconstructed and the identification route optimized further.

In contrast to the classification of this type of identification scheme as *monothetic* and *sequential* (Sneath and Sokal, 1973), this type of description really refers to the scheme's *method of construction* (totally automatic). The characters for the determinative set are selected one at a time (monothetically) and one after another (sequentially). Each selected character is optimized for partitioning or discriminating power in relation to the character selected previously. However, use of the determinative schemes is *simultaneous and polythetic* (totally automatic). In addition, the use of the schemes is programmed so that any results for

---

TABLE III. An example of use of an automated identification scheme (NEWARG) constructed from a traditional classificatory data set. This scheme is for 50 organisms versus 122 characters of the 144 organisms of subgroup 7 of Table I. The optimal test for character set for which binary results (1=present, 0=absent) are indicated under "Enter Test Results" is: catalase=0, ammonia from arginine=1, reduction of 0·1% methylene blue=0, mannitol=0, beta haemolysis=1, fermentative=1, grow on 10% bile blood agar=1, and sorbitol=0, or 01001110. This binary sequence of answers agrees with row number 79 on a $2^8=256$ row truth table. TESTAD is a subroutine for finding an answer from the first 10 possibilities in row 79. Test results for glycerol and the other tests listed are entered and the answer *Streptococcus agalactiae* is obtained. Note that the optimal test set for NEWARG is different from that for ARGPOC in Fig. 1. Optimal test sets differ as information is added or deleted from the data set. The test set is call optimal and not minimal because in a theoretical minimal set only six tests for characters would be required to partition 50 organisms ($2^6=64$)

```
B5700 TIME SHARING WITH DCP - 5/0
ENTER USER CODE, PLEASE - P16385
AND YOUR PASSWORD
████████
****WE ARE RUNNING ON DCPMCP.1369.  IF PROBLEMS, CALL 842-7167.****
09/12/73  8:55 AM.
GOOD MORNING, DR. RYPKA        YOU HAVE STATION 04


#
R NEWARG
 RUNNING


ENTER TEST RESULTS
?01001110
 01001110

POSSIBLE ORGANISMS ARE:

40823 STREPTOCOCCUS EQUISIMILIS
40825 STREPTOCOCCUS EQUI
40827 STREPTOCOCCUS SANGUIS
40828 S. ANGINOSUS, GROUP F
40829 S. ANGINOSUS, GROUP G
40830 S. AGALACTIAE
40882 STREPTOCOCCUS SP. 2
40884 STREPTOCOCCUS SP. 4
40885 STREPTOCOCCUS SP. 5
40886 STREPTOCOCCUS SP. 6

RUN TEST SERIES 7-C-79

GLYCEROL
HIPPURATE HYDROLYSIS
FIBRINOLYTIC
LACTOSE
RAFFINOSE
AESCULIN
TREHALOSE
 ARE ANY MORE TESTS DESIRED?
?NO


 END NEWARG 1.6 SEC.

R TESTAD
 RUNNING



ENTER SCHEME ID
?C-79
ENTER TEST RESULTS
?11000001
TESTS RESULTS EQUALS 40830 S. AGALACTIAE

ARE ANY MORE RESULTS REQUIRED
?NO


 END TESTAD .9 SEC.

BYE
 ON FOR  2 MIN, 23.0 SEC.
 C&E USE .7 SEC.
 EXECUTE 2.6 SEC.
 IO TIME 7.9 SEC.
 OFF AT   8:57 AM.
 GOODBYE P16385
09/12/73
```

162 E. W. Rypka

character variables may be entered simultaneously and all organisms having character states agreeing with the entered results are printed out in a submatrix with additional characters for testing in an attempt to discriminate the organisms in the subset. This type of scheme is operative daily via teletype and the Burroughs 5700 computer at the Lovelace Center for Health Sciences and the Scientific Laboratory Services of the State of New Mexico. In a similar manner this system has been used for transcontinental identification of microorganisms.

2. Reasons for Developing Other Methods

Data in Table I and Fig. 1 present corroborative evidence that microbial data sets are not acquired systematically and have a low information content. For these reasons the concept of continuous classification (Rypka, 1971) was developed as a simple, initial, and rather primitive way to trap data on-line at the source and to find rapidly if there is structure in the data set. The method is applicable particularly to situations where data generated by automatic measurement are trapped by interfacing the machine with an automatic data processing unit.

If a unique answer were not obtained by use of the methods shown in Table III and there were two or more organisms in the subset as determined by the first pattern of results, further testing for other characters may be done. If the answer involved nonseparable pairs of organisms—meaning further testing would not distinguish the possible pairs—the answer would be given as "nonseparable" and the likelihood (Fisher, 1934) or the normalized likelihood (Lapage et al., 1970) would not be calculated. Following suggestions of Payne (1963a, b) likelihood methods were developed for microbial identification (Rypka et al., 1967) but the success of the method was similar to that described by Dybowski and Franklin (1968). The limitation of the method is lack of information as indicated in Table I and the assignment of numerical values to peoples' linguistic habits. When this numerical assignment is done, *inferences somewhat beyond the data set are made*. Likelihood methods (Edwards, 1972) would be applicable at this time to excellent but small hard data sets such as those of Edwards and Ewing (1972). However, part of accuracy of identification involves the use of inclusive or exhaustive determinative schemes and until hard data are available for the occurrence of character states, likelihoods are not applicable optimally to large data sets. For these reasons just mentioned (lack of information and illogical interpretation of linguistic expression to numerical values) the methods of trapping data on-line, continuous truth table clustering, matrix learning, and self-correcting dynamic methods have been developed. These methods as discussed below allow for acquisition of a self-correcting hard data set for organism character variables to which methods of statistical inference more logically can be applied.

3. On-line Example

This example is an operative identification scheme and is a smaller part of subgroup 7 in Table I. The system works using a teletype in the laboratory and time-sharing on the Burroughs 5700 Computer. The subgroup contains 50 aerobic, Gram-positive cocci of medical importance including micrococci, staphylococci, sarcina, aerococcus, and streptococci. The character set for this subgroup is: catalase, ammonium production from arginine, ability to reduce 0·1% methylene blue, mannitol, beta haemolysis, fermentative, ability to grow on 10% bile blood agar, and sorbitol. These tests for characters are done and the results (present=1, absent=0) entered into the machine. The first response gives the row number of the binary sequence in a $2^8 = 256$ truth table and lists, if any, the possible organism or organisms (Rypka and Babb, 1970a).

If more than one answer is possible, a list of characters to test for is printed out. Then, a subroutine TESTAD is used, as indicated in Table III, to obtain the final identification. Various confidence levels for variable character variables may be used, for example, 99, 95, or 90% (Rypka and Babb, 1970a; Rypka *et al.*, 1967).

Data obtained on-line in this manner are extremely useful for acquiring phenotypic character data, antibiotic susceptibility data, epidemiology, and for studying host–parasite interaction (Rypka, 1971, 1972).

IDENTIFICATION SCHEMES CONSTRUCTED FROM
NUMERICAL CLASSIFICATORY DATA

Most practitioners of numerical taxonomy usually do not test the classes or clusters they have created for disjointness as an essential requirement of the classifications to be used for purposes of identification. Hester (Hester and Weeks, 1969, 1970; Hester and Rypka, 1973), in a study of the aerobic, Gram-positive bacilli, the flavobacteria, by single linkage numerical taxonomic methods (90 operational taxonomic units *vs* 104 character variables), found that identification schemes constructed from the cluster summaries yielded some differently designated clusters conjoint and not disjoint. He devised a different clustering methodology and was able to fulfil the basic tenet of classification— the creation of disjoint clusters or groups. The classifier must realize the importance of acquiring, if at all possible, a sufficient data set in order to be able to construct classifications so that all clusters and elements are disjoint. Failure to construct disjoint, exhaustive classes means future elements may not be identified correctly.

IDENTIFICATION SCHEMES CONSTRUCTED FROM
PRIMARY CLASSIFICATORY DATA

Primary classificatory data, such as that discussed by Gyllenberg (1964) under the designation of the "isolate sample method", are not difficult to obtain in laboratories dealing with high volume bacteriological work. Laboratories fulfilling this criterion include medical microbiological laboratories. As an illustration of a different method of using primary data, on-line data of enteric-resembling isolates were collected for purposes of demonstrating continuous classificatory data and the resulting identification schemes (Rypka and Lopez, 1973). Three types of matrices are considered in relation to primary data.

1. Frequency Classificatory Truth Table Matrix

The first type of matrix, the frequency classificatory truth table matrix (Rypka and Lopez, 1973), is that generated from data obtained at the source and the size of the data set is time dependent, the number of isolates (increasing with time) versus the number of characters for which character states have been determined. Table IV illustrates one such real data set in binary variables for aerobic, Gram-negative bacilli. A smaller example of how these data were collected, clustered, and the determinative scheme constructed is shown in Table VII (Rypka, 1972). Determinative schemes constructed from these primary data are based upon the frequencies of occurrence of isolates and these may vary temporally and geographically. Providing test sensitivities for detecting character states are adequate, the validity of the clusters increases as the number of isolates increases until sensitive and reasonably stable clusters may be obtained. The accuracy, rapidity, economy, stability and sensitivity of the determinative schemes, being classificatory data dependent, also increase with time.

2. Non-frequency Classificatory Truth Table Matrix

The second type of matrix, the non-frequency classificatory truth table matrix (Rypka, 1971), is shown in Table V. This matrix was obtained by taking each different row representing character states in Table IV and placing the different rows of 0–1 data in the order in which they occurred in the frequency classificatory truth table matrix. Identification schemes constructed from these types of classificatory data are more generalized and highly efficient. The characters selected for partitioning of elements in this data set are different to some extent and occur in a different order from those selected for the frequency classificatory truth table matrix. These differences in characters will be discussed after consideration of the third type of matrix.

FREQUENCY MATRIX
CHARACTERS

NON-FREQUENCY MATRIX
CHARACTERS

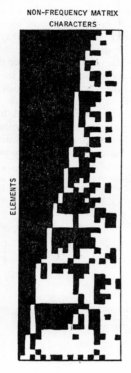

TABLE IV. This table (left) represents a primary data set with a frequency classificatory truth table matrix for over 100 aerobic, Gram-negative bacilli in subgroup 6, for 19 character variables recorded as present = 1 and absent = 0. In order, the columns are: glucose (acid), glucose (gas), rhamnose, sucrose, maltose, OF = fermentative, mannitol, fructose, catalase, oxidase, urease, motility, H_2S, indole citrate, lysine decarboxylase, pigment, nitrate reduction and morphology. All characters except xylose, are found in the optimal character set for ARGNER in Fig. 1. The data were clustered by continuous truth table methodology, and the data set was generated by Lopez (Rypka and Lopez, 1973). Nine characters selected optimally from the 19 characters partition the pairs of elements of the set according to curve 1 in Fig. 2

TABLE V. This table (right top) represents a primary data set with a non-frequency classificatory truth table matrix. Each different row pattern of binary variables in Table IV was placed in order to form this matrix. Nine characters (some different, some the same, but in different order than those selected in Table IV) partition the pairs of elements at the rate shown by curve 2 in Fig. 2. Note these data most closely approach the idealized theoretical dotted line curve

TABLE VI. Each cluster of apparent similarity in either Table IV or V was summarized (all ones = 1, all zeros = 0, and a mixture of ones and zeros = variable) to obtain this hierarchization matrix (right bottom) from the primary data set. Nine optimally selected characters from these data partition pairs of elements at the rate shown by curve 3 in Fig. 2. The solid square character states below the diagonal may be considered variables

HIERARCHIZATION MATRIX
CHARACTERS

TABLE VII. Data acquisition and continuous clustering by truth table classification (Rypka, 1972) are shown in a stepwise manner. Step 1: character data are obtained for four elements; Step 2: the characters are rearranged horizontally in descending order of the character with the largest number of ones; Step 3: each row of binary data is traced on the $2^5 = 32$ row truth table in Table II and the rows are rearranged vertically in ascending order of their truth table row number. The data are beginning to show structure. In step 4, binary character data are entered in the matrix in the order shown and in step 5 and step 6 rearranged horizontally by columns and vertically by rows as explained previously in steps 2 and 3. In step 7, the clusters of apparent similarity in step 6 matrix are summarized; this summarization procedure was explained in the legend of Table VI and the separatory value, $S = n_1 n_0$, for each character is calculated. In step 8, character 4 separates the most pairs of elements, then character 2 conditionally independent of 4 and lastly either 5 or 1 independent of 4 and 2. The truth table identification scheme is shown in step 9. Data in Table IV were clustered on-line by this procedure

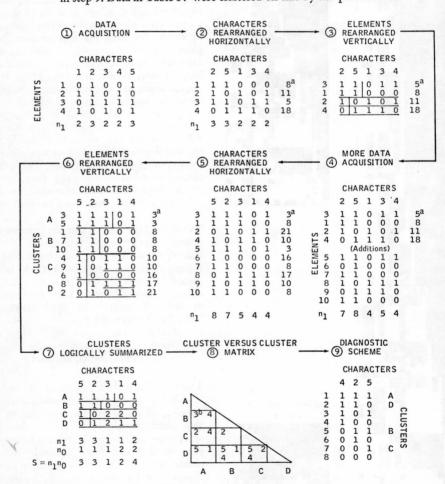

[a] Pattern Result Number from Table 2 (Two-valued Logic Matrix)
[b] Character 3 and 4 are different for A and B, Step 7.

3. Hierarchical Matrix

The third matrix, a hierarchization truth table matrix (Rypka and Lopez, 1973), is a summary of character states for isolates in the clusters in either of the first two classificatory truth table matrices. Examples of this type of matrix are given in Table VI. Vertical compression of the data has occurred and disjoint higher categories created. Identification schemes constructed from hierarchization truth table matrices may have different optimal partitioning characters from the previous two matrices.

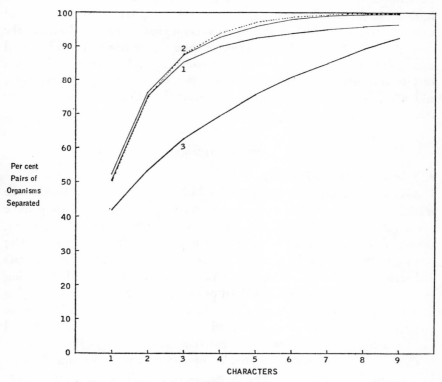

1. Frequency matrix, optimum selection of characters for separation.
2. Non-frequency matrix, optimum selection of characters for separation.
3. Hierarchization matrix, optimum selection of characters for separation.
   Characters selected vary for each matrix and the order of similar characters is different.
4. Theoretical curve. $S = [1 - V^{-k}] / [1 - V^{-c}]$. Section VI. C. Equation 4.

FIG. 2. Optimal rates of separation of pairs of elements by characters selected for the primary data set shown in Tables IV, V, VI (Rypka and Lopez, 1973). The dotted curve is the theoretical curve shown by use of the formula given after (4) above. The character order and characters are not exactly the same for each curve.

The character test sets and rate of separation of pairs of elements for the identification schemes for the three types of matrices are shown in Fig. 2. The importance of optimal selection of characters for partitioning elements cannot be overemphasized when exhaustive or inclusive schemes are used. Economy of schemes in terms of material and time required for identification are important factors from a cost center viewpoint in addition to the accuracy and rapidity.

As emphasized previously, the concepts of continuous classification and identification are considered as initial methods of trapping data of attribute states of elements, objects, and events and of finding structure in data sets as we learn in time. As data sets become larger we are surpassing our manual and mental capacities to find readily an underlying structure based upon *all* the data—the expert notwithstanding. An advantage of these simple horizontal and vertical matrix rearrangements of character states and elements in columns and rows is that structure may be revealed less abstractly than in paired comparison methods, and clusters of apparent similarity are formed with human and machine recognizable patterns.

GENERAL THEORY

1. *Inhaltlich* vs *Axiomatic Theory*

Having gained experience with the microbial data set in binary variables, the methodological application was extended to other data sets such as drug abuse studies, facial recognition, cloud clustering, human blood biochemical variables, medical diagnosis, variables associated with human error (Kowalsky *et al.*, 1974; Rypka, 1972) and chess playing. These data sets consisted of mixed variable types and scale conversion of variables was necessary. This led to variables being studied in the context of many-valued or polyvalued logic and attempting to develop a formalized axiomatic theory of classification and identification in contrast to the concrete axiomatic methods (*inhaltlich*) based upon empirical information that have been used in the past (Hilbert and Bernays, 1934, 1939).

There are a considerable number of operative systems for automatically handling medical information.* Such systems, both small and large, do adequate work of sorting and formatting medical data for readability, storage, and retrieval. Few if any systems, however, actually attempt to build the microbial

---

* "Clinical laboratory computer systems. A comprehensive evaluation" (1971). J. Lloyd Johnson Associates. 1500 Skokie Building, Northbrook, Illinois.

Communication and computer devices for pathologists (1971). Western Regional Joint Meeting. College of American Pathologists and the California Society of Pathologists. St Francis Hotel, Union Square, San Francisco. September 25–26.

or the medical concept in time and to use the acquired data set to construct disjoint classifications and optimal diagnostic routes. These data could be used in matrix learning or concept learning (Hunt, 1962; Hunt *et al.*, 1966) in a manner similar to the way in which microbial information has been handled. The data set synthesis in truth table classification and determination, utilizing many-valued logic instead of 0–1 data, finds more structure in the data sets and assuredly helps to establish a more logical and less intuitive approach to critical decision-making. The interplay of filling missing bits in the data set matrix and the expanding of these matrices both horizontally (character variables) and vertically (elements, objects, or events) has solid practical and theoretical implications. A simple method based upon many-valued truth tables is presented below in a general manner to show ostensibly how the system operates. These methods of matrix learning (Steinbuch and Piske, 1963) and similar interactive methods discussed by Shortliffe *et al.* (1973) really are not involved with nor should they be considered examples of artificial intelligence. Decision rules, programming, and disambiguation of factors relevant to diagnosis hardly are determined artificially at this time.

2. Measurement and Scale Conversion of Variables

(a) Measurement. Description of elements by states of their character variables could involve a method of measurement which may be simply an observation of color or shape to some highly objective measurement for detecting micro-quantities of a substance. Churchman (1959) has commented that we do not know why we measure and that measuring does "violence" to reality. In any case, propositions about character states describing elements in classificatory data sets range from qualitative statements that may involve preciseness or ambiguity and vagueness to quantitative statements which assign numbers to characters which describe elements, objects, and events—this leads, in turn, to higher levels of abstraction. Real data sets, particularly in biology, consist of mixed character variable types. Rarely do we stop the thought process, or turn thought back upon itself, in an attempt to understand how decisions are reached on the basis of using mixed character variable types. Decision is involved in the act of measurement and a great deal more study is required concerning the fundamentals of measurement although excellent work has been done in this area (Anderberg, 1972; Churchman, 1959; Ellis, 1968; Torgerson, 1958).

(b) Scale conversion of variables. Variables assigned number values of measurement may be classified according to size of the range set as binary or dichotomous, discrete, and continuous. Also, variables may be classified according to the measurement scale as nominal, ordinal, interval, and ratio (Anderberg, 1972;

Ellis, 1968). Examples of these variable types and their cross classification is shown in Table VIII. This table is used with the permission of M. Anderberg (1972). A crucial matter in the development of the general theory that follows is scale conversion of measurements of character variables into states of logic. For

TABLE VIII. Cross–classification of variables and examples are shown. Problems of measurement and scaling data are discussed in the text. This table is used with the permission of M. Anderberg (from Anderberg, 1972)

VARIABLES

Scale of Measurement \ Range of the Variable	Continuous: May assume an uncountably infinite number of values	Discrete: May assume a finite (or at most countably infinite) number of values	Binary: May assume only two values
Ratio: If $x_A > x_B$, A is x_A / x_B times greater than B and $x_A - x_B$ units greater than B.	Temperature in °K; weight; height; age	Counts such as numbers of children, hospitals, cars	Unit price of soft drinks in vending machines bottles or cups: 10¢ cans: 15¢
Interval: If $x_A > x_B$, A is $x_A - x_B$ units greater than B.	Temperature in °C; specific gravity	Serial numbers; TV channel numbers	How many wives do you have? (assuming the only legal answers are 0 or 1)
Ordinal: Either $x_A > x_B$, $x_A = x_B$ or $x_A < x_B$.	Human judgements of texture, brightness, sound intensity	Military rank; models in a line of cars (Ford, Mercury, Lincoln); wide, medium, narrow	Tall-short; good-bad; big-small; wide-narrow
Nominal: Either $x_A = x_B$ or $x_A \neq x_B$	Absurd – requires an uncountably infinite number of distinct classes	Eye color; place of birth; favorite actor	Yes-no; present-absent; dead-alive; on-off; true-false

(left margin label: VARIABLES)

Used with permission of the author :

example, the measurement for each character variable may be partitioned into five parts, designated states or values of logic, and symbolized by an integer (0, 1, 2, 3, 4) or by a letter (A, B, C, D, E). These may be called *poly-* or *n-chotomous variables*. This constitutes a partitioning of all scales of measurement of variables in the data set into the same number of states. However, variables in data sets may be partitioned into different logic states: for example, one variable may be partitioned into eight states, another into two states. These considerations of different types of variables in mixed data sets closely agree with actual human treatment of real world information. Scale conversion of variables in data sets is really a matter of judgment best reserved for the investigator who is familiar with the set and who may want to apply multi-state clustering methodology to his data.

3. General Identification Equation

(a) *Clustering.* Identification schemes may be constructed directly from un-clustered data. Clustering, however, may indicate structure in a data set and by summarizing the variables in clusters the data set may be reduced considerably. The method of truth table classification has been illustrated for binary variables in Tables IV, V and VI with the method for data acquisition shown in Table VII. A similar application has been made using n-valued characters in Table IX. In

TABLE IX. Explanation of the clustering methodology is given in the text. These clustering techniques are extensions of continuous classification in binary variables to n-valued logic. Real world data sets used in decision making usually involve mixed variable types. Many-valued logic clustering and identification can handle this type of data.

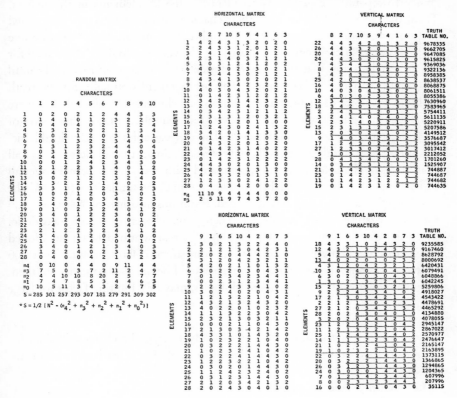

the top two matrices, clustering was done by first rearranging the columns of the random matrix in descending order of the columns with the largest number of the highest values of logic; logic value $4>3>2>1>0$. Then the rows of elements with these patterns of states of variables were rearranged vertically in descending order of the truth table numbers of a matrix with $5^{10}=9765625$ rows of unique patterns to obtain the clusters shown in the vertical matrix. Calculation of the truth table row number for each matrix row is automatically done; the clustering is done automatically. These clusters are summarized, the matrix reduced, and an identification scheme constructed from the summarized matrix. This would be a hierarchization matrix similar to that shown in Table VI.

A second method of clustering follows closely the method of selecting characters for the identification. Clustering by this methodology helps explain the method of construction of schemes and the general identification equation discussed in the next two sections. In the bottom two matrices in Table IX, the characters of the random matrix are rearranged in descending order of their ability to separate or partition—to identify—the 28 elements in the matrix. Then, the rows of elements with their character patterns are again rearranged vertically in descending order of their truth table number. Note there are five larger clusters of nearly equal size created by the values of logic of character 9 for the 28 elements and that these larger clusters are subclustered repeatedly until 28 unique elements are recognized. This arrangement optimally fits the truth table for purposes of identification. These two clustering methodologies were developed to be used with large data sets, for example, a matrix dimension of 4000 elements, objects, or events vs 400 character variables in 2–22 valued logic. A limiting factor with some coefficients of association is that an element over all of its states of character variables is compared with every other element over all of its states of character variables and the number of similarities (n_s) and the number of differences (n_d) is determined. From this information $S=n_s/n_s+n_d$* for each organism pair is calculated. Then, some clustering technique is done. With the matrix dimension (4000×400) of a real world data set mentioned above, this would involve 7 998 000 pair comparisons (4000 (3999)/2) over 400 character variables before clustering even would begin. Pair comparison, then, in some numerical taxonomic methodologies, may be a limiting factor for large data sets.

(b) *Construction of the identification scheme.* By tracing the character values for character 9, vertical matrix bottom right in Table IX, of each element on the

---

* s = similarity (a coefficient of association).

TABLE X. Three truth tables, $5^1 = 5$, $5^2 = 25$, and $5^3 = 125$, for identification of the 28 elements in Table IX. Pairs nonseparable by the first three selected characters are separable in the subscheme. The empirical and theoretical separation of pairs is shown in Fig. 3. To sort elements into the truth tables, the character logic values for the character 9, 9-1, and 9-1-6 were traced over the proper truth table and the element number placed after the row where the elements character data agreed

TRUTH TABLES IN FIVE-VALUED LOGIC

$5^1 = 5$

1	4	18, 12, 15, 13
2	3	4, 10, 6, 1
3	2	15, 9, 17, 2, 27, 28, 3
4	1	23, 11, 25, 14, 21, 19
5	0	22, 20, 26, 24, 7, 8, 16

$5^2 = 25$

1	44	
2	43	12, 18
3	42	5
4	41	
5	40	13
6	34	
7	33	
8	32	
9	31	4
10	30	10, 6, 1
11	24	
12	23	15
13	22	9
14	21	17, 2, 27
15	20	28, 3
16	14	
17	13	
18	12	23, 11
19	11	25, 14
20	10	21, 19
21	04	
22	03	22, 21, 26, 24
23	02	
24	01	7
25	00	8, 16

$5^3 = 125$

1	444		26	344		51	244		76	144		101	044	
2	443		27	343		52	243		77	143		102	043	
3	442		28	342		53	242		78	142		103	042	
4	441		29	341		54	241		79	141		104	041	
5	440		30	340		55	240		80	140		105	040	
6	434		31	334		56	234		81	134		106	034	
7	433	18	32	333		57	233		82	133		107	033	
8	432	12	33	332		58	232	15	83	132		108	032	22, 20
9	431		34	331		59	231		84	131		109	031	26
10	430		35	330		60	230		85	130		110	030	24
11	424		36	324		61	224		86	124		111	024	
12	423		37	323		62	223		87	123		112	023	
13	422		38	322		63	222	9	88	122	23	113	022	
14	421		39	321		64	221		89	121	11	114	021	
15	420	5	40	320		65	220		90	120		115	020	
16	414		41	314		66	214		91	114		116	014	
17	413		42	313		67	213	17	92	113		117	013	
18	412		43	312	4	68	212	2, 27	93	112	25	118	012	7
19	411		44	311		69	211		94	111	14	119	011	
20	410		45	310		70	210		95	110		120	010	
21	404		46	304		71	204		96	104		121	004	
22	403		47	303		72	203		97	103		122	003	
23	402	13	48	302	10, 6, 1	73	202	28, 3	98	102	21, 19	123	002	8
24	401		49	301		74	201		99	101		124	001	16
25	400		50	300		75	200		100	100		125	000	

PAIRS TO SEPARATE

10-6
10-1
6-1
2-27
28-3
21-19
22-20

SUBSCHEME CHARACTERS

	5	10	4
10	4	0	2
6	2	0	3
1	1	1	3
11	0	3	0
2	1	3	4
28	4	3	0
3	0	4	4
21	3	2	4
19	3	2	2
22	2	4	1
20	2	2	1

$5^1 = 5$ row truth table in Table X, the elements are sorted as indicated. The separatory value for character 9 is (Rypka *et al.*, 1967):

$$S = 1/2[G^2 - (n_4^2 + n_3^2 + n_2^2 + n_1^2 + n_0^2)] \qquad (1)$$

or
$$S = 1/2[28^2 - (4^2 + 4^2 + 7^2 + 6^2 + 7^2)]$$
$$= 309.$$

The separatory value for characters 9 and 1 are found by using the $5^2 = 25$ size truth table and the equation in the form

$$S = 1/2[G^2 - (n_{44}^2 + n_{43}^2 \ldots n_{00}^2)] \qquad (2)$$

or
$$S = 1/2[28^2 - (n_{43}^2 + n_{42}^2 + n_{40}^2 + n_{31}^2 \ldots n_{00}^2)]$$
$$= 1/2[28^2 - (2^2 + 1^2 + 1^2 \ldots 2^2)]$$
$$= 360.$$

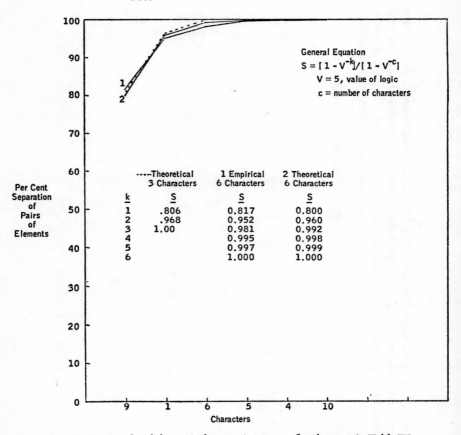

General Equation
$$S = [1 - V^{-k}] / [1 - V^{-c}]$$
$V = 5$, value of logic
c = number of characters

	----Theoretical 3 Characters	1 Empirical 6 Characters	2 Theoretical 6 Characters
$\underline{k}$	$\underline{S}$	$\underline{S}$	$\underline{S}$
1	.806	0.817	0.800
2	.968	0.952	0.960
3	1.00	0.981	0.992
4		0.995	0.998
5		0.997	0.999
6		1.000	1.000

Per Cent Separation of Pairs of Elements

Characters

FIG. 3. Empirical and theoretical separation curves for elements in Table IX.

The separatory value for characters 9, 1 and 6 using the $5^3 = 125$ value truth table in Table X is

$$S = 1/2[G^2 - (n_{444}^2 + n_{443}^2 \cdots n_{000}^2)] \qquad (3)$$

$$= 371.$$

The minimal character set required to separate 28 elements is $T_{\min} = \log_v G$ (G = elements in group, v = value of logic, $T_{\min}$ = minimal number of characters to result in theoretical separation) (Rypka, 1972) and $\log_5 28 = 3$ (rounded to next highest integer). The number of pairs of elements to separate is $G(G-1)/2 = 378$. The first three characters, 9, 1 and 6 will separate 371 of these pairs or $371/378 \times 100 = 98 \cdot 1\%$ of the pairs. Character 5 will separate remaining pairs 10–6, 10–1, 6–1, 2–27, and 28–3 ($376/378 = 99 \cdot 5\%$); character 4 will separate pair 21–19 ($377/378 = 99 \cdot 7\%$) and character 10 separates pair 22–20 ($378/378 = 100\%$). The rate of separation of pairs by these six characters which were selected optimally is shown in Fig. 3. These selection methods are automated.

(c) *General identification equation.* Development of concepts thus far have dealt with practical consideration of clustering random classificatory data sets several different ways and of constructing identification schemes from the frequency, non-frequency, and the hierarchical classificatory truth table matrices. Consideration of truth tables, such as that shown in Table XI, indicates that each character variable partitions the same number of elements in the element versus element matrix and according to equations (1) and (2). The theoretical rate of separation of elements for each character, conditionally independent of the previously selected character in Table XI, would be

Character 1: 250 pairs

Character 1 and 2: 250 + 50 pairs

or in general (Rypka, 1972)

$$S = (1 - v^{-k})/(1 - v^{-c}) \qquad (4)$$

S = separatory value for the vth value of logic, the kth character, and c = the number of characters. The theoretical curve is shown for three characters and five valued logic in Fig. 3.

E. W. Rypka

TABLE XI. An illustration of the theoretically ideal case of many-valued logic is shown. Five-valued logic and two characters are used so the truth table, shown at the left, has $5^2 = 25$ patterns of two, five-valued variables which theoretically represent unique character state combinations of 25 elements. In the element versus element matrix, every possible pair of elements in the truth table is compared over all characters. When character states or values are different the elements of the pair are separable or disjoint. In this ideal case all pairs of elements are separable. There are $G(G-1)/2 = (28 \times 27)/2 = 378$ pairs of elements to partition. The top right half of the table is a mirror image of the bottom left half. Real world data sets rearranged to agree with truth tables cluster the data and create optimal partitioning of elements. Theoretical curves for a $5^3 = 125$ truth table used in Table X are shown in Fig. 3

SUMMARY

The objective of this work was the development of a methodology for continuous classification and continuous identification in an attempt to increase the information content in microbiology. Classifications are dynamic and tentative simply because all elements to be classified are usually unavailable at one time for study. The living forms present an ever changing pattern of characteristics. Any method that continuously would classify the living forms at a high level of similarity and yet create disjoint clusters so that identification would be more definitive would improve greatly the biological classificatory and determinative data set. The method is cyclic and the validity of identification is contingent upon the validity of classification. The tacit assertion is made throughout the discussion that classification is considered an activity of inductive inference and identification an activity of deductive inference; that is, classification and identification are inverse mental activities. The method is analogous to finding an instantaneous rate of change of a function. In the present work the objective was to find an instantaneous correlation of phenotypic character position and pattern of elements with time for purpose of continuous classification and from these classifications to construct simultaneously usable identification schemes.

ACKNOWLEDGEMENTS

I wish to acknowledge thanks to Pamela McKay Martin for typing the manuscript; to Pauli Paulikonis, R. Speeker Rypka, B. A. Hanson for illustrations; to Robert Babb for many years of cooperative effort in programming; to Royce Fletcher and Michael Anderberg for discussionary assistance, and mathematical development for eqn (4). NIGMS 16385, generated data in Table I and Fig. 1, Lovelace Center for Health Sciences, and Roche Diagnostics (Division of Hoffman–LaRoche) provided travel funds.

REFERENCES

ANDERBERG, M. R. (1972). "Cluster analysis for applications". Office of the Assistant for Study Support. Kirtland Air Force Base, New Mexico. Report No. OAS–TR–72–1, AD 738301.

ASHBY, W. R. (1970). "An Introduction to Cybernetics". Chapman and Hall, London.

BLACKITH, R. E. and REYMENT, R. A. (1971). "Multivariate Morphometrics". Academic Press, London and New York.

BREED, R. S., MURRAY, E. G. D. and SMITH, H. R., eds (1957). "Bergey's Manual of Determinative bacteriology" (7th edition). Williams and Wilkins Co., Baltimore.

CHURCHMAN, C. W. (1959). Why measure? *In* "Measurement: Definition and Theories". (C. W. Churchman and P. Ratoosh, eds). Wiley, New York.

COLE, A. J. (1969). "Numerical Taxonomy". Academic Press, London and New York.

DUDA, R. O. and HART, P. E. (1973). "Pattern Classification and Scene Analysis". Wiley–Interscience, New York.

DYBOWSKI, W. and FRANKLIN, O. A. (1968). Conditional probability and the identification of bacteria: A pilot study. *J. gen. Microbiol.* **54**, 215–229.

EDWARDS, A. W. F. (1972). "Likelihood". Cambridge University Press.

EDWARDS, P. R. and EWING, W. H. (1972). "Identification of Enterobacteriaceae". (3rd edition). Burgess, Minneapolis.

ELLIS, B. (1968). "Basic Concepts of Measurement". Cambridge University Press.

FIRSCHEIN, O., FISCHLER, M. A., COLES, L. S. and TENENBAUM, J. M. (1974). Intelligent machines are on the way. *IEEE Spectrum* **11**(7), 41–48.

FISHER, R. A. (1934). Two new properties of mathematical likelihood. *Proc. R. Soc.* A **144**, 285–307.

GOSE, E. E. (1969). Introduction to biological and mechanical pattern recognition. *In* "Methodologies of Pattern Recognition". (S. Watanabe, ed.), pp. 203–252. Academic Press, New York and London.

GRUSSER, O. J. and KLINKE, R., eds (1971). "Pattern Recognition in Biological and Technical Systems". Springer Verlag, Berlin.

GYLLENBERG, H. G. (1963). A general method for deriving determinative schemes for random collection of microbial isolates. *Ann. Acad. Sci. fenn.* A. IV. Biologica **69**, 5–22.

GYLLENBERG, H. G. (1964). An approach to numerical description of microbial populations. *Ann. Acad. Sci. fenn.* A. IV. Biologica **81**, 3–23.

HESTER, D. J. and RYPKA, E. W. (1973). Identification of numerical taxonomical clusters of bacteria. *Abstracts of the Annual Meeting of the American Society for Microbiology.* Miami. 35.

HESTER, D. J. and WEEKS, O. B. (1969). Taxonometric study of the genus *Brevibacterium* Breed. *Bact. Proc.* 19.

HESTER, D. J. and WEEKS, O. B. (1970). Taxonometric studies of Gram-positive pigmented bacteria. *Abstracts Xth International Conference on Microbiology.* Mexico. D.F. 192.

HILBERT, D. and BERNAYS, P. (1934, 1939). "Grundlagen der Mathematik", Volumes 1 and 2. Quoted in Kneebone, G.T. (1963). "Mathematical Logic and the Foundations of Mathematics. An Introductory Survey". Van Nostrand, London.

HUNT E. B. (1962). "Concept Learning. An Information Processing Problem", Wiley, New York.

HUNT, E. B., MARIN, J. and STONE, P. J. (1966). "Experiments in Induction". Academic Press, New York and London.

IEEE (1972). Workshop on pattern recognition applied to man and his environment. Sponsored by IEEE Computer Society Technical Committee on Pattern Recognition. Organizing Committee: C. K. Chow, P. F. Lambert, G. Nagy, and E. A. Parrish. The Homestead, Hot Springs, Virginia.

JARDIN, N. and SIBSON, R. (1971). "Mathematical Taxonomy". Wiley, London.

KANAL, L. and CHANDRASEKARAN, B. (1969). Recognition, machine "recognition", and statistical approaches. *In* "Methodologies of Pattern Recognition" (S Watanabe, ed.), pp. 317–332. Academic Press, New York and London.

KOWALSKY, N. B., MASTERS, R. L., STONE, R. B., BABCOCK, G. L. and RYPKA, E. W. (1974). An analysis of pilot error-related aircraft accidents. 1–72. Contract No. NAS4-1931. NASA CR-2444. National Aeronautics and Space Administration. Washington, D.C.

LAPAGE, S. P., BASCOMB, S., WILLCOX, W. R. and CURTIS, M. A. (1970). Computer identification of bacteria. *In* "Automation, Mechanization and Data Handling in Microbiology". (A. Baillie and R. J. Gilbert, eds), pp. 1–22. Academic Press, London and New York.

MEISEL, W. S. (1972). "Computer-oriented Approaches to Pattern Recognition". Academic Press, New York and London.

MINSKY, M. (1968). "Semantic Information Processing". The MIT Press, Massachusetts Institute of Technology, Cambridge, Massachusetts.

MINSKY, M. and PAPERT, S. (1969). "Perceptrons". The MIT Press, Massachusetts Institute of Technology, Cambridge, Massachusetts.

MICHENER, C. D. and SOKAL, R. R. (1957). A quantitative approach to a problem in classification. *Evolution* **11**, 130–162.

PAYNE, L. C. (1963a). New computing techniques in medicine. *Proc. R. Soc. Med.* **56**, 560–562.

PAYNE, L. C. (1963b). Toward medical automation. *Wld med. Electronics* **2**, 6–11.

RYPKA, E. W. (1971). Truth table classification and identification. *Space Life Sci.* **3**, 135–156.

RYPKA, E. W. (1972). Pattern recognition as a method of studying host–parasite interaction. 1972 International Conference on Cybernetics and Society. IEEE Systems. Man and Cybernetics Society and American Society for Cybernetics. October. Washington, D.C.

RYPKA, E. W. and BABB, R. (1970a). Automatic construction and use of an identification scheme. *Med. Res. Engng* **9**, 9–19.

RYPKA, E. W. and BABB, R. (1970b). Automatic construction and use of inclusive identification schemes. Abstracts. 191. Xth International Congress for Microbiology. August. Mexico, D.F.

RYPKA, E. W. and LOPEZ, L. (1973). Taxonomy: application of truth table classification and the general identification equation to on–line isolates. *Abstracts of the Annual Meeting of the American Society for Microbiology, Miami.*

RYPKA, E. W., BOWEN, L. G. and BABB, R. (1967). A model for the identification of bacteria. *J. gen. Microbiol.* **46**, 407–424.

SHORTLIFFE, E. H., AXLINE, S. G., BUCHANAN, B. G., MORGAN, T. C. and COHEN, S. N. (1973). An artificial intelligence program to advise physicians regarding antimicrobial therapy. *Comp. biomed. Res.* **6**, 544–560.

SNEATH, P. H. A. (1957a). Some thoughts on bacterial classification. *J. gen. Microbiol.* **17,** 184–200.

SNEATH, P. H. A. (1957b). The application of computers to taxonomy. *J. gen. Microbiol.* **17**, 201–226.

SNEATH, P. H. A. and SOKAL, R. R. (1973). "Numerical Taxonomy". Freeman, San Francisco.

SOKAL, R. R. and MICHENER, C. D. (1958). A statistical method for evaluating systematic relationships. *Univ. Kansas Sci. Bull.* **38**, 1409–1438.

SOKAL, R. R. and SNEATH, P. H. A. (1963). "Principles of Numerical Taxonomy". Freeman, San Francisco.

STEINBUCH, K. and PISKE, V. A. W. (1963). Learning matrices and their application. *Trans. Inst. Elec. Electron. Eng. Elec. Comp.* **EC-12**, 846–862.

TORGERSON, W. S. (1958). "Theory and Methods of Scaling". Wiley, New York.
WATANABE, S. (1969). Pattern recognition as an inductive process, *In* "Methodologies of Pattern Recognition" (S. Watanabe, ed.), pp. 221–234. Academic Press, New York and London.

12 | An On-line Identification Program

R. J. PANKHURST* and R. R. AITCHISON

Department of Botany, University of Cambridge, England

Abstract: A conversational computer program is described which allows identification of specimens by working in a question-and-answer manner at a computer keyboard. This program was prepared for teaching and demonstration purposes, and demands the exclusive use of a small computer. The operation of the program is described with a detailed example.

Key Words and Phrases: identification, on-line computing, diagnostic characters, Rubus

INTRODUCTION

The program operates by allowing the user to choose a character of his specimen and describe it to the machine. The machine then compares this information with its reference data, and eliminates any taxa which disagree. This process continues step by step until, ideally, exactly one possible identification remains. This is similar in principle to the use of a diagnostic key, except that one is free to use any characters in any sequence. A better comparison would be with a polyclave (see review by Morse, 1971), where punched cards or something equivalent are used.

EXAMPLE

The program begins by typing a title (see Appendix I). This particular example applies to *Rubus fruticosus* microspecies in the county of Cambridgeshire. The machine types READY whenever it is idle waiting to be given a new command.

We continue by asking for a LIST of CHARACTERS. This list contains all characters which can be used to distinguish species at this moment (diagnostic characters). In other words, all characters whose values are identical are not included, since they cannot be used to eliminate any of the species. These are called redundant characters. Notice that characters with missing values may be

* Present address: British Museum (Natural History), London.

Systematics Association Special Volume No. 7, "Biological Identification with Computers", edited by R. J. Pankhurst, 1975, pp. 181–194. Academic Press, London and New York.

useful, but that taxa for which the value is missing will not be eliminated by using such characters. We then select a character, for which we are prepared to give details, and type CHARACTER, then, in response to WHICH CHARACTER? we give PANICLE PRICKLES CURVE. The option between values CURVED or STRAIGHT is offered, and we select STRAIGHT by typing the value number (36). The program then tells us how many species are still possible. After specifying the FLOWERS COLOUR, a further reduction is obtained.

At this point, we type VERIFY. This is an optional command, and its effect will be seen in a moment. Meanwhile, LIST CHARACTERS is repeated. One can see how the contents of this list have altered. PANICLE PRICKLES CURVE and FLOWERS COLOUR have gone, because they have already been used. LEAVES PINNATE OR PALMATE has gone because it is now redundant. GLANDS ON PLANT has now appeared because it used to have missing values, but for the set of species now in question, it is fully scored. When we enter PETALS WIDTH the machine asks this time if we really mean it. This is the effect of VERIFY, and it acts as a check. There was no mistake, and we type YES, and the character is accepted.

In a forgetful moment, we next tried to use the PANICLE PRICKLES CURVE character again, and the machine rejects this. The next character, LEAVES FELT, shows what happens when NO is typed at a request for verification. After this, the machine actually types the names of the remaining species, since this list is now fairly short. The command LIST SPECIES LAST gives the species list before the last character was entered, so we can see what species were then eliminated. Notice that one name, RUBUS SCABROSUS, appears twice. This is because it appears several times in the data, as it is rather variable.

The next character, STEM HABIT, is given to show how we can change our minds after a character has been given. With the command BACK, the most recent character can be erased, and we return to the previous species list. We now give some more characters, and the VERIFY command, given a second time, cancels the effect of the first. With the character STEM PRUINA, we typed the wrong value altogether, and the machine objected with a message. A little later, we typed LIST while forgetting to say what we wanted listed, and the machine has again objected. The character SEPALS DIRECTION brings us down to one species only, RUBUS PURPUREICAULIS, and we have finished the identification of this specimen. The command RETURN clears the machine ready for the next specimen.

A detailed specification of all the possible commands is given in Appendix II.

PROGRAM DESIGN

An on-line program must either operate as part of a time-sharing system, or else make use of a computer of moderate size which is temporarily dedicated to this purpose alone. The latter choice was made on the grounds of convenience. A time-sharing system was also available, but there were administrative barriers in the way of its use.

The machine in question was a DEC PDP 11/45 with about 24 000 words of main storage, an input and output typewriter, and a random-access disc store. This is large enough for the purpose, but not so powerful that to dedicate it temporarily to on-line identification alone would seem wasteful. The disc store is not directly needed for the program, but it accommodates a FORTRAN compiler with a disc operating system (DOS). The program was written in standard FORTRAN, and although it was mainly developed on a separate large machine (IBM 370/165) in a batch processing mode, operating the PDP11 without the disc system would have been very inconvenient.

The program operates on the same input data as those required for the companion key-generating program (Pankhurst, 1973). In other words, the same pack of cards can be used either for automatic key-construction or for on-line identification. In these data, each character and value is given an integer number for reference. For example, FLOWERS COLOUR might be character number 4, and the relevant values might be WHITE, PALE PINK OR PINK, DARK PINK with values 1, 2 and 3 respectively. Each species is described by a string of numbers, representing each character in turn. For example, if the 4th number is 3, this indicates PETALS DARK PINK.

The program is constructed so that it runs in a continuous cycle. It is either waiting for a user command to appear, or else is in the process of carrying out the last one, during which time it will not accept other commands. Most commands begin with a unique four-letter sequence, and these are looked up in an array of permitted commands. The index of a legal command in this array is used as the index of a computed GO TO, in order to branch to the right part of the program to carry out the command.

Besides the array in which the taxonomic data matrix is held, two other arrays are of central importance. These are used to remember which characters, and which species, are applicable at any one time. Initially, all species and all diagnostic characters are applicable. As soon as a character is specified, that character becomes inapplicable (i.e. used). All species which do not agree with the value of this character also become inapplicable. Further, all characters which then turn out to be redundant at this point are made inapplicable. When the number of applicable species becomes reduced to one, then that species is the result, and the identification is complete. The BACK command is programmed by making a copy of the previously applicable characters and species before a new character is entered, and restoring these when required.

The program, as it stands, accepts only qualitative data. There is, however, no essential difficulty about expanding it to cater for quantitative data as well. It is a simple matter to switch from one set of input data to another, and a selection of different applications is kept in store on a magnetic tape. The typewriter is a little slow and somewhat noisy, and a computer-driven display would be better in many ways. A display might be more expensive, however, and it does not give a written record of an operating session.

As a general principle, the operation of the program and the messages it produces have been planned to give most importance to the convenience of the user. We believe this is essential in order to attract and hold the interest of many taxonomists who are, quite rightly, not interested in playing with computers for their own sake, and who will demand that their path be made easy.

OTHER WORK

The program described by Boughey *et al.* (1968) is more comprehensive and sophisticated than ours. It operates on a time-shared system, and includes the use of a computer driven display. It allows for both quantitative and qualitative characters, and can carry out identification by several alternative procedures. The work of Goodall (1968) is related to that of Boughey.

The program of Morse (1974) is roughly similar to the one described here. He calls it a "polythetic" program, since it has one special and very useful feature. The user is allowed to state that he will allow species to disagree with the specimen by a certain number of characters. All species which disagree up to this limiting number of characters are retained, but when the names are printed, the number of disagreements (if any) is printed as well. This allows the user to make some mistakes in the description of the specimen, or to have an atypical specimen, and still get some correct results. The rate at which the identification proceeds will be slower, however. This idea is not especially difficult to program. It shows that there is not any particularly clear distinction between elimination (key) methods and matching methods.

ADVANTAGES AND DISADVANTAGES

The on-line approach is more flexible than that of a printed key, especially when the polythetic version is used, since one is free to choose any characters in any sequence. Strictly, one can only choose diagnostic characters, but the machine can be made to inform the user which these are. The on-line method also has a great deal of emotional appeal.

On the other hand, some of the disadvantages of the key method are retained. A wrong observation may lead the user permanently astray, and there is no numerical confidence measure in the result. Also, the method is severely dependent on the availability and accessibility of a suitable machine which might render it entirely impractical for some potential users. It is certainly not suitable for use in the field.

CONCLUSIONS

This program was developed for the sake of completeness as a part of a project on identification by computer, and is not intended to be particularly novel. It has proved very valuable for purposes of teaching and demonstrating, and the response from students has been very encouraging. A program of this nature is recommended for its teaching value rather than its utility.

REFERENCES

BOUGHEY, A. S., BRIDGES, K. W. and IKEDA, A. G. (1968). An automated biological identification key. Museum of Systematic Biology, University of California, Irvine, Research series No. 2.

GOODALL, D. W. (1968). Identification by computer. *BioScience* **18**(6), 485–488.

MORSE, L. E. (1971). Specimen identification and key construction with time-sharing computers. *Taxon* **20**(2/3), 269–282.

MORSE, L. E. (1974). Computer programs for specimen identification, key construction, and description printing using taxonomic data matrices. Biological Series, Publications of the Museum of Michigan State University **5**, 1–128.

PANKHURST, R. J. (1973). Documentation for the key generating program. Internal report, *but see also*: Botanical keys generated by computer. *Watsonia* **8**, 357–368 (1971).

Appendix I

Example of a run with the on-line identification program.

Note. All lines beginning * are typed by the user, all the rest are typed by the machine.

RUBUS IN CAMBRIDGESHIRE

```
    READY

*LIST
*CHARACTERS

    FLOWERS COLOUR
    GLANDS ON PLANT
    STEM HAIRS
    STYLES COLOUR
    FLOWERS DIAMETER
    ACICLES ON PLANT
    STEM PRICKLES NO.
    LEAVES LEAFLET NO.
    LEAVES FELT
    STEM ANGLES
    PETALS WIDTH
    PANICLE PRICKLES CURVE
    LEAVES PINNATE OR PALMATE
    LEAVES LACINIATE
    STEM PRICKLES REL LENGTH
    READY

*CHARACTER

    WHICH CHARACTER?

*PANICLE PRICKLES CURVE

    35   CURVED
    36   STRAIGHT

*36

    31 SPECIES REMAIN

    READY

*CHARACTER

    WHICH CHARACTER?
```

```
*FLOWERS COLOUR

     1  WHITE
     2  PALE PINK OR PINK
     3  DEEP PINK

*2

   19 SPECIES REMAIN

 READY

*VERIFY

 READY

*LIST
*CHARACTERS

 GLANDS ON PLANT
 STEM HAIRS
 SEPALS DIRECTION
 STYLES COLOUR
 FLOWERS DIAMETER
 ACICLES ON PLANT
 STEM PRICKLES NO.
 LEAVES LEAFLET NO.
 LEAVES FELT
 LEAVES PEDATE OR DIGITATE
 STEM ANGLES
 PETALS WIDTH
 FILAMENTS REL STYLES
 STEM PRICKLES REL LENGTH
 READY

*CHARACTER

 WHICH CHARACTER?

*PETALS WIDTH

    31  MORE THAN HALF AS LONG AS BROAD
    32  ABOUT HALF AS LONG AS BROAD

*31

 DO YOU MEAN PETALS WIDTH MORE THAN HALF AS LONG AS BROAD?

*YES

   13 SPECIES REMAIN

 READY

*CHARACTER

 WHICH CHARACTER?

*PANICLE PRICKLES CURVE
```

G

```
CHARACTER ALREADY USED, CHOOSE AGAIN
READY

*CHARACTER

WHICH CHARACTER?

*LEAVES FELT

    22   GREY OR WHITE FELTED BENEATH
    23   NOT FELTED OR GREEN FELTED BENEATH

*23

DO YOU MEAN LEAVES FELT NOT FELTED OR GREEN FELTED BENEATH?

*NO

READY

*CHARACTER

WHICH CHARACTER?

*LEAVES FELT

    22   GREY OR WHITE FELTED BENEATH
    23   NOT FELTED OR GREEN FELTED BENEATH

*22

DO YOU MEAN LEAVES FELT GREY OR WHITE FELTED BENEATH?

*YES

WELL DONE, YOU HAVE ONLY 9 POSSIBILITIES LEFT

    RUBUS CARDIOPHYLLUS
    RUBUS CONJUNGENS
    RUBUS CRINIGER
    RUBUS DISCERPTUS
    RUBUS POLYANTHEMUS
    RUBUS PROCERUS
    RUBUS PURPUREICAULIS
    RUBUS SCABROSUS
    RUBUS VESTITUS
    READY

*LIST
*SPECIES LAST

    RUBUS BABINGTONIANUS
    RUBUS CARDIOPHYLLUS
    RUBUS CONJUNGENS
    RUBUS CRINIGER
    RUBUS DISCERPTUS
    RUBUS POLYANTHEMUS
    RUBUS PROCERUS
    RUBUS PURPUREICAULIS
    RUBUS SCABROSUS
```

```
RUBUS SCABROSUS
RUBUS TUBERCULATUS
RUBUS VESTITUS
RUBUS CAESIUS
READY
```

*LIST
*CHARACTERS

```
STEM HAIRS
SEPALS DIRECTION
CARPELS HAIRS
GLANDS ON PLANT
FLOWERS DIAMETER
STEM PRICKLES NO.
LEAVES PEDATE OR DIGITATE
STEM ANGLES
FILAMENTS REL STYLES
STEM HABIT
READY
```

*CHARACTER
WHICH CHARACTER?

*STEM HABIT

```
56   HIGH ARCHING
57   ARCHED-PROCUMBENT
```

* 56

DO YOU MEAN STEM HABIT HIGH ARCHING?

*YES

WELL DONE, YOU HAVE ONLY 2 POSSIBILITIES LEFT

```
RUBUS CARDIOPHYLLUS
RUBUS CONJUNGENS
READY
```

*BACK

THE LAST CHARACTER USED WAS LEAVES FELT
READY

*LIST
*SPECIES

```
RUBUS CARDIOPHYLLUS
RUBUS CONJUNGENS
RUBUS CRINIGER
RUBUS DISCERPTUS
RUBUS POLYANTHEMUS
RUBUS PROCERUS
RUBUS PURPUREICAULIS
RUBUS SCABROSUS
RUBUS VESTITUS
READY
```

```
*CHARACTER

 WHICH CHARACTER?

*GLANDS ON PLANT

     4  FEW OR NONE
     5  NUMEROUS

*?

 DO YOU MEAN GLANDS ON PLANT NOT CONSPICUOUSLY HAIRY?

*NO

 READY
*CHARACTER

 WHICH CHARACTER?

*GLANDS ON PLANT

     4  FEW OR NONE
     5  NUMEROUS

*5

 DO YOU MEAN GLANDS ON PLANT NUMEROUS?

*YES

 WELL DONE, YOU HAVE ONLY 4 POSSIBILITIES LEFT

 RUBUS CRINIGER
 RUBUS DISCERPTUS
 RUBUS PURPUREICAULIS
 RUBUS SCABROSUS
 READY

*VERIFY

 READY
*LIST
*CHARACTERS

 STEM HAIRS
 SEPALS DIRECTION
 FLOWERS DIAMETER
 ACICLES ON PLANT
 STEM PRICKLES NO.
 LEAVES PEDATE OR DIGITATE
 STEM PRUINA
 STEM ANGLES
 READY

*CHARACTER

 WHICH CHARACTER?
```

*STEM PRUINA

   27   PRUINOSE
   28   EPRUINOSE

*26

BAD LUCK-SPECIMEN DOES NOT AGREE WITH DATA

THE FOLLOWING SPECIES REMAINED LAST TIME

RUBUS CRINIGER
RUBUS DISCERPTUS
RUBUS PURPUREICAULIS
RUBUS SCABROSUS
READY

*BACK

THE LAST CHARACTER USED WAS GLANDS ON PLANT
READY

*CHARACTER

WHICH CHARACTER?
*LEAVES SIZE

CHARACTER REDUNDANT, CHOOSE AGAIN

READY

*CHARACTER

WHICH CHARACTER?

*STEM PRICKLES NO.

   18   UP TO 15 PER INTERNODE
   19   MORE THAN 15 PER INTERNODE

*19

WELL DONE, YOU HAVE ONLY 2 POSSIBILITIES LEFT

RUBUS PURPUREICAULIS
RUBUS SCABROSUS
READY

*LIST

INPUT UNRECOGNISABLE

READY

*LIST
*CHARACTERS

SEPALS DIRECTION

```
    STEM ANGLES
    READY

*CHARACTER

    WHICH CHARACTER?
*SEPALS DIRECTION

      8   REFLEXED
     33   ERECT OR CLASPING

*8

    WELL DONE, YOU HAVE ONLY 1 POSSIBILITIES LEFT

    RUBUS PURPUREICAULIS
    READY

*RETURN
    READY

*CHARACTER
    WHICH CHARACTER?

*FLOWERS COLOUR
```

.
.
.
.
.
.
.
.
.
.
.
.
.
.
.
.
.
.
.
.
.
.

Appendix II

Commands for On-line Identification Program

At the start of the program the title of the selected data is printed followed by a READY. After READY the operator has a choice of 6 commands.

At the completion of each command the machine types READY and the program returns to this point.

(a) LIST

On the next line either: (i) CHARACTERS—gives a list of useful characters;
                (ii) OBJECTS—gives a list of objects still available;
                (iii) OBJECTS LAST—gives a list of objects still available before the last character was added.

(b) CHARACTER—to input a chosen character.

Response from machine WHICH CHARACTER?

The operator then types the selected character exactly as it appears in the character list as obtained from (a) (i).

The machine responds with a list of the possible values for that character each preceded by a number. The operator selects the value appropriate to his specimen and types its number.

The machine then responds in one of three ways:

  (i) n OBJECTS REMAIN
  (ii) WELL DONE! ONLY m POSSIBILITIES LEFT
     followed by a list of these, where m is less than 10.
  (iii) BAD LUCK! SPECIMEN DOES NOT AGREE WITH DATA.
     This can mean that the specimen is not included in the data, or that the operator has typed a wrong value.

Errors occur when typing in the character name and the machine may respond with

  (i) CHARACTER UNRECOGNIZABLE, CHOOSE AGAIN
     This is probably caused by an error in typing the character.
  (ii) CHARACTER REDUNDANT, CHOOSE AGAIN
     The character is of no use for separating the remaining objects.
  (iii) CHARACTER ALREADY USED, CHOOSE AGAIN
     In all these cases it is often helpful to list the characters as in (a).

(c) VERIFY

This is an optional command which enables the operator to check that the character-value pair chosen is the one intended. After the value number has been typed VERIFY produces DO YOU MEAN Character-value?

The operator has the option to type YES or NO. If YES is typed, the value is accepted. If NO is typed, the character is ignored and READY is printed. One must then select a character once more.

(d) BACK

This enables the operator to go back to the previous character and the last one is ignored. This is useful when the last character produced the message SPECIMEN DOES NOT AGREE WITH DATA. Typing BACK will now ignore the last character used and save the operator starting again at the beginning. This command cannot be used more than once at a time.

(e) RETURN

This returns program control to the READY position and unsets the VERIFY command. This must be used at the completion of identification of each specimen. The program is then ready for another identification session.

(f) FINISH

To terminate the program.

Note: A mis-typed character or characters in a line is destroyed by pressing a "rub-out" key an appropriate number of times. This convenience was supplied by the operating system.

Taxonomic Data for Identification

13 | A Generalized Descriptive Data Bank as a Basis for Computer-assisted Identification*

STANWYN G. SHETLER

Smithsonian Institution, Washington, D.C., U.S.A.

Abstract: The generalized approach to systems design and development for taxonomic data banking has been emphasized from the outset of the Flora North America Program. With a generalized descriptive data base and a generalized processing software, the user can perform identifications by matching characteristics of an unknown specimen against the stored characteristics of known species. Such a generalized approach places stringent requirements on the data base. The descriptive data must be truly comparative from species to species, regardless of taxonomic group. The system must be able to diagnose any one species relative to all others and not merely in relation to certain near neighbours. Vocabulary standardization is of paramount importance. For this, nested relationships and synonymy of terms must be dealt with realistically, and a minimum set of characters must be described for all species. The FNA System operates with a predefined, classified vocabulary. A given character may be described by any number of character-states, and each character-state may be qualified by any number of the approved modifiers.

Key Words and Phrases: computer-assisted description, computer-assisted identification, data bank, data retrieval, flora, Flora North America, GIS, identification, information system, key-forming, morphological description

INTRODUCTION

The move to computerization should not be the automatic cause for setting aside all previous convention; often conventional procedure provides the best algorithm for programming the computer. The step to computers is a major undertaking in itself, and the taxonomist hardly needs to take upon himself gratuitously the added burden of pioneering entirely new conventions for his science. At the same time, the mere simulation of a conventional procedure is not necessarily desirable nor the most effective or economical way to use the machine. The time-honored dichotomous key, for example, is designed for use

* Flora North America Report No. 76.

Systematics Association Special Volume No. 7, "Biological Identification with Computers", edited by R. J. Pankhurst, 1975, pp. 197–235. Academic Press, London and New York.

in conventional printed form. Nothing is gained by computerizing such keys, although the computer can be used advantageously in construction of conventional keys. On the other hand, the computer offers many new possibilities in the direct use of so-called "polyclaves" or "multi-entry (multi-access) keys". Thus, if the computer is to be applied to the tasks of identification and key construction, care must be taken to design a system of computerization that will result in truly new capability and not in mere mimicry by machine of conventional practice.

From the outset the Flora North America (FNA) Program has emphasized a generalized approach to systems design and development for taxonomic data banking. The term "generalized", as used by FNA, is meant to convey more or less the notion of "multi-purpose". It is not intended to suggest a sketchy or vaguely designed system, as the term tends to mean in many contexts. Each new purpose imposes its own set of constraints on the design and thereby limits the extent to which the system can serve any other function. Unlike a single-purpose system, which can be designed to perform its function optimally, a multi-purpose system must be designed with generalized functions that optimize the system as a whole without necessarily optimizing any one of its applications. Likewise, the stored data (i.e. the data bank) must be sufficiently general in content and format to serve any or all of the predefined purposes.

Generalized systems are much more difficult to design and build, therefore, than tailored systems. The designer, assuming a common data base, must chart beforehand the input and output of each proposed application, and he must foresee and specify in detail each major product and service of the system. Too often it is assumed that generalized systems are systems designed to absorb *miscellaneous* data and then to perform on the principle of serendipity, producing desired but previously undefined and unanticipated results if and when they are needed. Although any designer certainly hopes for some serendipitous payoffs, one never could develop a system, generalized or tailored, without predefining at least one major purpose (application). A data bank requires a plan exactly as a book requires a plan, and gathering data for a multi-purpose data bank may be likened to collecting data in a single filing system to support the writing of several books. There is no logical way to begin collecting data either for books or data banks before the purposes are defined.

The FNA goal has been to design and build a system capable of producing a complete Flora of approximately conventional form directly from a single, comprehensive data bank. In other words, the object has been to create a data bank that is sufficiently generalized in structure, format, and content to supply all output required for the Flora, and to develop a system that is sufficiently generalized in function to perform all processing and report-generating necessary to produce the Flora directly by machine from the data bank. This Flora is to be printed either from camera copy off the standard line-printer of the computer or from tape copy coded to supply a computer-driven type-setting and printing process. Final copy from the system must be acceptable as to syntax, grammar, punctuation, and typography. These are severe constraints even for a tailored, single-purpose system.

As part of this concept, therefore, it was planned initially that the system should be capable of constructing on command the conventional dichotomous keys required in the printed Flora directly from the elements stored in the common data bank. Furthermore, the system was to be capable of supporting an on-line interactive identification service as the need demanded. Consequently, early effort was focused on designing a system that could meet the requirements for dichotomous key construction and on-line identification. This led to an early, mutualistic association with Larry E. Morse, who has become widely known for his innovative program package for interactive key construction and identification (Morse, 1974). The proposed FNA system posed a large-scale practical problem in the kind of computer application conceived by Morse. The urgent impetus of FNA stimulated Morse to enlarge his concept and to develop his programs as rapidly as possible. While Morse provided FNA with a ready-made pilot project with which to test the concept of a system focused on identification and key construction, FNA provided Morse with a ready-made testing ground and significant intellectual and financial support for his approach. For a period of years, FNA investigated, tested, and promoted the Morse system vigorously. Through this collaboration Morse has made an indelible contribution to the dynamic evolution of the present FNA System.

Gradually it became apparent that our preoccupation with computerizing the taxonomic functions of identification and key construction was diverting our attention from the central requirements of a truly generalized system for biological data retrieval and Flora-writing, namely, a generalized data base and a generalized processing software for creating, maintaining, and querying data files. Above all, our system had to enable us to pose a variety of queries, to select subsets of data using Boolean logic and complex selection criteria, and then to generate a report in whatever format we wished to specify at output time, concatenating records and elements at will to produce the desired format. The system also had to be capable of handling large data bases and had to provide convenient, economical means of building large files rapidly. The last criterion, for example, virtually precluded any software that permitted only on-line creation and maintenance of files.

Analysis of the larger FNA requirements and extensive experience with Morse's program package combined to dictate that in the FNA System identification and key construction should be uncoupled from description. Apart from the fact that special systems such as Morse's system, despite their obvious utility within the proper limits, are not sufficiently developed yet to be practical on a flora-wide basis, the FNA Editorial Committee was forced to conclude reluctantly after much investigation, debate, and actual testing that it could not have everything in one system at this stage. It would have to choose between a system designed primarily to store and permute prefabricated descriptive phrases composed to be diagnostic for the purpose of forming keys and identifying plants, and a system designed primarily to store and permute the elemental data describing the plants. Given this choice, the Committee could only select the latter, more generalized alternative, because it wanted to be able to deal in the traditional descriptive currency of the taxonomist and above all to be able to query the data base on demand for subsets of data defined by any number of user-selected criteria. It was decided that the specialists would construct their own keys in the time-honored conventional way, because, given the peculiar nature of keys, on the whole this still is the most efficient way for the specialist to produce them. The finished keys would be stored line for line in a printer file on magnetic tape for intercalation at the proper places in the Flora as it is being printed out by computer for publication. The Committee hoped, of course, that in time a specialized identification system such as Morse's could be linked with the more generalized FNA System and possibly could feed from the same data base or at least from files within the same data bank.

MAJOR CONSIDERATIONS IN DEVELOPING A GENERALIZED SYSTEM

1. Specialized Nature of Key-Forming Process

The dichotomous key is an ingenious intellectual invention for comparing and contrasting organisms explicitly with the least possible information. Key construction is a highly specialized and personalized process, perhaps as much an art as a science. The statements that might describe an organism form a virtually limitless universe of values. Sifting this universe for just the right values is a purely human function. Fashioning a necessary and sufficient small number of paired diagnostic phrases so as to maximize the discrimination and minimize the words always requires an expert, whether or not the computer ultimately is used at some point in the formulation and printing of the key. The phrases must distil tersely the essence of the distinctions, epitomizing the common features while optimizing the separations.

Most variation in nature occurs in continua, and the taxonomist's task in preparing a dichotomous key is to reduce the continuum to discrete values, insofar as possible to binary choices between absolute conditions. It is his job to depict the variability in blacks and whites and to eliminate the intermediate grays at least for the purposes of the key. Such analysis, evaluation, and decision-making hardly can be relegated to the computer, no matter how sophisticated the programming. Without reducing the natural continua to discrete values, there is little hope of producing truly workable keys.

By contrast with key construction, generalized description calls for as much emphasis on the similarities as the differences, and the taxonomist's responsibility is to be scrupulous in describing the continuous as well as the discontinuous variation—to point out the grays as well as the blacks and whites.

Thus key-construction and description tend to be antithetical processes, and it is difficult if not impossible to conceive of a single data base and set of procedures that would optimize both functions, whether or not the machine is used. The currency of description is the elemental term or datum, whereas the currency of key construction is the carefully constructed phrase mentioned earlier, which is precast for a particular comparison or type of comparison in a particular context or class of contexts. The key-construction data base is a synthetic data base of a higher order, therefore, than the generalized descriptive data base.

There basically are two possible approaches to using the same computerized data base for both key-forming and descriptive purposes, neither of which appears at this stage to be feasible on a large practical scale.

The first is to attempt description and generalized descriptive data retrieval from a key-construction data base built up of comparative phrases. The specialized phrases as used in a key seldom are meaningful by themselves, however, because outside the exact context of the key for which they were framed they lose most if not all of their comparative message. Their information content usually is so relative that they have little or no absolute meaning. The likelihood of writing a computer program that could read a miscellany of phrases, break them into component units, edit the units, and re-synthesize an accurate and coherent description seems extremely slim. The slightest shades of difference in phraseology and the mundane idiosyncrasies of punctuation and syntax conspire to thwart the formulation of logical and consistent rules for automatically splicing phrases together into an acceptable description. On the other hand, the formulation of predictable and trustworthy query routines for generalized data retrieval seems equally unlikely. Record scanning certainly is not the answer. Even if these were not insurmountable problems, the fact remains that such phrases in the aggregate become highly repetitive, with the same basic terms recurring numerous times in different permutations within and among taxonomic groups. It soon becomes apparent to anyone who works with key-forming phrases in the computer that the storing of data as paired, contrasting statements constitutes an awkward, bulky, and highly inefficient way to develop a general data retrieval system for diagnostic and descriptive information. One never could store an exhaustive list of phrases in the machine even if one could imagine finite limits to couplet-making and could conceive of compiling the list. Indeed it is questionable what the value is of storing key phrases in the computer for general permutation and retrieval, inasmuch as they must be printed out as they were read in, that is they are for display only.

The second alternative is to attempt key-forming by machine from a generalized descriptive data base. Ideally, of course, one would hope to synthesize keys automatically in this way from the elemental data. Even if one is willing to assume the existence of a more or less exhaustive data base, however, it seems doubtful if not improbable that an algorithm ever can be designed and programmed to simulate all of the steps in the intellectual process that an expert taxonomist goes through in defining the comparisons and concatenating the terms to construct a key.

2. Diagnostic vs Comparative Description

Describing organisms is potentially a limitless task, and for practical purposes most authors of Floras, Faunas, monographs, revisions, and other descriptive works concentrate on the obvious differences. The plants or animals are described relative to each other, rather than according to some objective general standard, which might permit general comparison and correlation.

Because the diagnostic characters differ greatly from group to group, the descriptions provided in a Flora, for example, rarely are comparable from one family or genus to another. Usually, description goes little beyond a succinct statement of the "key" or diagnostic features. This extremely relative, diagnostic character of description is seen in its purest form in the dichotomous key. Even the vocabulary changes, as in the ferns where the leaf becomes the frond, the petiole the stipe, etc. As a consequence, the statements about a particular species or group of species in a key or description make sense only in the context of the particular group involved. In the context of the plant kingdom at large, many of the statements will have no objective, universal meaning even throughout a group as large as a family, not to mention a major subdivision, e.g. the vascular plants.

Some examples from actual keys will serve as illustrations. A couplet in one of the numerous keys of Bailey's "Manual of Cultivated Plants" (1949) poses, for instance, this delightfully relative choice:

Plant prized in dry state, the stems curling up in a ball when dry, expanding
    when put in water.
Plant prized in its growing state.

This hardly is a characteristic susceptible of standard definition and general application, and even the *Selaginella* expert, using this key as intended, is likely to have difficulty interpreting so relative a feature.

This more orthodox example, selected from the same work, is reasonably representative of keys in general and appears at first sight to stand by itself in or out of context:

Resin ducts marginal.
Resin ducts internal.

One quickly surmises, perhaps, that this couplet belongs to a key to some group of coniferous plants, but to which part of the plant does it apply? Stem (trunk)? Leaf? In fact, the couplet compares branchlets between certain species of *Abies*, but elsewhere in the same key the same character pair is applied to leaves. Here, then, is a typical example of a character pair with explicit diagnostic meaning in the precise context of a particular key to a particular group of plants, but without intrinsic comparative meaning in the wider context of plants generally.

In conventional descriptive works, the taxonomic classification itself is used to indicate implicitly the extent to which plants or animals have characteristics in common. The very taxonomic position of an organism is an implicit statement of properties, which can be enumerated explicitly if and when necessary. Two plants or animals placed in the same category are understood to have more or less characteristics in common, and the lower the taxonomic rank of that category the more features they can be expected to share until they share everything except individual variation. Thus, to say that a given genus of plants belongs to a given family of plants is to convey implicitly a whole array of characteristics shared by all known members of the family. For the expert, this is a highly economical means of conveying information, conserving space in the mind as well as on paper or in computer storage, because a simple name is used to represent a whole catalog of common features, and he is free to emphasize only the important differences—the truly diagnostic features.

Too often, however, reliance on the taxonomic classification as an information system leads to unchecked and meaningless assumptions. Organisms frequently are classified on the basis of relatively few obvious characteristics, and once the act of classification has been performed there is a tendency to anticipate or presume that all other features, if examined, will be found to be consistent with the taxonomic position as determined. Indeed, the act of naming itself tends too easily to take on exaggerated significance, because often it satisfies the practical needs to the point of removing all incentives for further study and verification (Raven *et al.*, 1971). With the whole pattern of scholarship and practice in taxonomy focusing on means of classifying and identifying organisms, the incentive to collect patently non-diagnostic data merely for the sake of consistency and completeness is minimal or lacking altogether in most studies. As a consequence, comparisons among large numbers of organisms generally are difficult if not impossible to make for lack of truly comparative data (see also Pankhurst, this volume, Chapter 14).

The computer now provides the technical means for making complex comparisons and correlations rapidly and consistently, but the potential of the machine never can be realized until certain historical inadequacies in the diagnostic system of taxonomic description are rectified. Without a rational system for amassing a data base of truly comparative descriptions, it is impossible to design a rational system for retrieving, comparing, and correlating such descriptive data by machine, whether for purposes of identification, description, or query response. If, for example, one has an unknown plant specimen that has marginal resin canals in the leaves, one cannot search among known plant species for other examples of such resin canals, whether by hand or by machine, unless every species in the data base already has been scored in a standard way for presence or absence of marginal resin canals in the leaves. This would presuppose a standard definition and usage of the terms "leaf", "canal", "resin", and "marginal".

The keys to comparability and machine-processability are standardization and consistency. Without applying a standard set of characters and a standard descriptive terminology throughout the range of plant species being described, there is no hope of collecting truly comparable data for all species. The characters and terms must be classified and defined insofar as possible on an absolute basis so that they will have an unambiguous application and an intrinsic, uniform meaning apart from any particular context. Each character must be unique and self-defining, either by virtue of unique position in a classification or unique terminology. Thus, if the character "resin duct" can apply to more than one plant part, the plant part must be specified in proper hierarchical relationship (see below) to the plant as a whole and in *each* instance where the character is described.

Standardization is a necessary but not a sufficient condition of data comparability. Equally necessary is the *consistent* recording of characters across the range of specimens or taxa. If a character is recorded for one specimen or taxon, it should be recorded for all specimens or taxa if at all possible, whether or not the character is of diagnostic importance. On the other hand, there is little point in recording a character for one specimen or taxon if there is no chance of recording it in reasonable time for most or all specimens or taxa. This is to say that the characters to be described should be limited to that set for which there is a reasonable prospect of obtaining character-state values in all cases. Otherwise the problem of missing data becomes bothersome at the least and critical at the worst by undercutting the precision and predictability of the retrieval procedure. One cannot rely, for instance, on the results of a search for all species with marginal resin ducts in the leaves if in fact this character was recorded for only 50–75% of the species in the data base. It is just as important to record the absence of a feature as its presence, and this must be done in a conscious, affirmative way so as to distinguish between the condition of absence and the condition of no knowledge. The terminology for absence or degrees of absence should be parallel to the terminology for the state or states of presence. In other words, there should be no missing data as such. Each character of each taxon within the circumscription of the data base should be assigned a character-state from a predefined list of presence/absence alternatives, and when the information to assign such a value does not exist, a null value (to indicate lack of information) should fill the blank.

Finally, the computer cannot be expected to "know" the relationships and shared features implicit in the systematic position of an organism, as signified by its taxonomic name. These must be spelled out consistently and completely in each record, or there must be links to files where they are explicitly stated. The expert will know, for example, that members of the plant family Asteraceae (Compositae) are characterized by the inflorescence type termed a "head". For the computer to know this, the inflorescence type must be stated for each species, or each record must be chained in some manner to a common record for the family in which the inflorescence type and other universally shared features are described once for all members.

3. Character Nesting
Taxonomic descriptions typically are written in a highly telegraphic, technical shorthand that only the expert can interpret reliably. Many terms implicitly signify character nesting and other relationships that only the taxonomist will perceive. The term "rhizome scaly", for example, tells the taxonomist, among other things, that he is dealing with a plant part, that this part is a modified, underground stem lacking in chlorophyll, and that the type of vesture on the

surface of this underground stem is squamose (scaly). The nested relationships might be expanded into the following hierarchy:

```
PLANT
  STEM
    STEM, ABOVE-GROUND
    STEM, UNDERGROUND
      TYPE—rhizome
      SURFACE
        SURFACE, ADAXIAL
          VESTITURE
            TYPE—squamose
        SURFACE, ABAXIAL
          VESTITURE
            TYPE—squamose
```

For certain purposes no doubt even more levels would be needed in this nested character classification, but the example serves to illustrate the complexities that arise quickly when one attempts to express explicitly the full hierarchical chain of relationships implicit in a simple statement like "rhizome scaly".

For a machine-processable data base all of these implicit relationships must be made explicit if the searching and sorting capacity is to be logical and generalized. The term "rhizome scaly" is accessible only to queries that hinge on the two words "rhizome" and "scaly" and their combination. The same information when elaborated in a hierarchy as above is accessible to a great variety of questions hinging on any of the words or combinations of words in this hierarchy. For example:

Which plants have stems?

Which plants have underground stems?

Which plants have rhizomes?

Which plants have stems that are scaly underground on both the adaxial and abaxial surfaces?

etc.

4. Synonymy

Synonyms, which so enrich spoken and written language, are the bane of information transfer systems, particularly mechanized or computerized systems. In the strictest sense, there is no such thing as a synonym. Two words or terms may convey the same sense for all practical purposes, but inevitably there are contexts in which one term is preferable to the other because of a shade of difference in meaning. The trained and experienced mind can comprehend the shades of difference and match the terms to the contexts or perceive underlying unity or identity when several basically synonymous words are used interchangeably to vary the style and add color. Not so the machine!

Taxonomists are past masters at exploiting the shades of difference in the words of the language. Over the years they have adopted or manufactured one term after another to describe the features of plants and animals as precisely as possible. The shades of difference become especially apparent in the descriptive terminology that has evolved to describe such features as vestiture or pubescence, where qualitative terms have been made to serve in the place of measurements and statistics to describe quantitative variables. Perhaps the major cause of proliferation of terms is the taxonomist's concern that his descriptive language be neutral with respect to evolutionary connotations. Thus to avoid inferring homology without sound basis taxonomists and other biologists have resorted to numerous, often pedantic terminological circumlocutions that tend to get in the way even of the scientists themselves when they attempt to compare organisms across groups in a simple, straightforward way. Whole vocabularies of specialized terms evolve around a particular family or other plant group, as, for example, in the case of the higher plant families Asteraceae, Orchidaceae, and Poaceae (Gramineae). These special vocabularies are necessary to cope with the extreme modification of certain groups of organisms, to be sure, but there are many examples of superfluous terms that would have been better left uncoined. Anyone attempting to create a machine data base must wrestle with the excess baggage of superfluous terms as well as with the large nucleus of essential terms and try in some manner to simplify the overall vocabulary of description. Otherwise the problem of description becomes totally quantitative, whether terms or numbers are used, and stretches to infinity.

To complicate the matter still more, some taxonomists prefer to use the classical Latin or Greek terms or their legitimate, immediate derivatives, while others prefer to describe their organisms in the modern vernacular. Consequently, in any language in which a significant amount of scientific description is done, a dual vocabulary of sorts develops in which the classical terms have their counterparts in the native language, e.g. suberous *vs* corky, squamose *vs* scaly, lamina *vs* blade.

Variations in spelling and format represent still another type of synonymy. Here, for example, are several format variations of the country name UNITED STATES OF AMERICA:

UNITED STATES
UNITED STATES.
   UNITED STATES (note left-hand margin variation)

U.S.A.
U. S. A. (note spacing variation)
U.S.
U. S.
USA

To any educated person reading these several variants, the meaning is the same, but to an indexer, whether man or machine, these designations of country are all different. As far as the machine is concerned, when it is asked to match or to sort, they represent eight different countries.

Isolated cases of synonymy of this type or other types, if they can be delimited, can be accommodated in a data bank by incorporating translation tables in the computer program(s). Contingency programming is not the general answer, however, to the general problem of synonymy. The only general answer is a standardized vocabulary. The limitations of programming strategies are obvious. Even for a simple case like the one above, it is virtually impossible to account for all the variations in format and spelling that in practice might be encountered. To provide this type of translation processing for every single term in the data base is unthinkable. Even if it were possible to program the computer to make every conceivable translation, storing the equivalency tables would take astronomical space, and processing speed would be degraded to the point of making the cost of processing astronomical. Furthermore, although translation tables work for demand searching, they do not solve the problem of sorting the data into indices and other listings where it is necessary to have like terms appear side by side. In the above example, the abbreviations and full words would not alphabetize together, and the form of punctuation and spacing similarly would affect sorting order.

The designer of a machine-searchable data base has no alternative, therefore, but to standardize his vocabularies and formats rigidly from the outset by means of authoritative lists (authority files). Syntax, grammar, word order, spacing, spelling, abbreviation, capitalization (if upper and lower case are differentiated by the machine), and punctuation must be *identical* for all terms and statements intended to convey identical information or to sort (alphabetize) at the same place in an index or catalog. This means that the order of modifiers relative to nouns must be uniform throughout and that, therefore, the designer must decide in advance which part of a particular term, if it comprises two or more words, is to be emphasized and given the initial position for classificatory or sorting purposes. For the term BOREAL AQUATIC PLANTS, for example, the designer has at least three major options, depending on which of the three words is to be given primacy:

BOREAL AQUATIC PLANTS
AQUATIC PLANTS, BOREAL
PLANTS, BOREAL AQUATIC

At least 14 legitimate and predictable variants of this three-word term, including the above three, are possible without getting into abbreviations. Reversing the order of AQUATIC AND BOREAL, for example, has little effect on the meaning of the term except to a mechanical classifier or indexer such as the computer. Some systems designers insist on eliminating all unnecessary spaces and punctuation and might require that a term of this type be rendered without punctuation, regardless of word order. Here, then, are the other 11 obvious variants:

AQUATIC PLANTS,BOREAL
AQUATIC PLANTS BOREAL
PLANTS, BOREAL AQUATIC
PLANTS BOREAL AQUATIC
AQUATIC BOREAL PLANTS
BOREAL PLANTS, AQUATIC
BOREAL PLANTS,AQUATIC
BOREAL PLANTS AQUATIC
PLANTS, AQUATIC BOREAL
PLANTS,AQUATIC BOREAL
PLANTS AQUATIC BOREAL

5. Report-Generating Requirements

A data bank must be organized and created with no less care than is required to write and produce a book. In some respects the preparation of a data bank is far more demanding. The goals to be achieved and the limits within which they are to be accomplished must be predefined in scrupulous detail. Otherwise data will be accumulated aimlessly. The actual print format must be defined completely at the outset, also, because the data must be input and stored in ready-to-concatenate-and-print form.

In effect, therefore, the data banker must set his own type word by word and data element by data element at input, because at output he does not have an opportunity to make individual changes. At output, only *classes* of changes that can be made systematically and mechanically by program are possible. For one reason or another, this solution usually proves to have severe limitations. Idiosyncrasies in the data, such as individual misspellings or misplaced commas, must be corrected individually. If all such corrections are not made one by one *before* the final report is generated, the report will have to be generated repeatedly until the data meet acceptable standards, much as successive galley- or page-proofs might be run off. With the best of efforts, vetting a large, complex manuscript or data base of typographical mistakes tends to be an asymptotic function of the number of drafts or proofs. Thus, there is no more efficient and economical way to vet data than to write them down or input them correctly in the first place. After input, corrective updating always costs more time and money, and such costs quickly can rise out of sight when a data base is large and is being accumulated over a long period of time.

If the designer of a data bank is to derive the major benefits of computerization, he must organize his data in a sufficiently generalized form that the elemental units, like atoms, can be compounded at will into whatever larger units are desired in the output. Much forethought must go into the format as well as the organization of the fields or building blocks of data in the system. Questions such as whether to use the nominal or adjectival form of a term and whether to use the singular or the plural must be decided on a consistent basis before any data are input to the system. As already indicated, unless the print format is worked out in minute detail in advance by the designer, it is likely that many data will have to be re-entered, perhaps repeatedly, or that computer programs will have to be written endlessly to rectify the data for outputting. Such programs tend to be dangerous because they often have unpredictable side effects that domino through the data base.

Finally, a major reason for using the computer is to have the capability at the end of a data-banking operation not only to produce a catalog or similar synopsis of the data base, but also to sort various data fields into indices that cross-reference the catalog. If such automatic indexing cannot be done with a minimum of further editing, then much of the original data-banking effort has been in vain.

6. Software Selection

In a recent paper, Krauss (1973a) defined a generalized information processing system and discussed the important reasons for using such a software package in biological data banking. To the criteria for assessing generalized systems that she outlined should be added capability for handling hierarchical data structures and capability for querying two or more files simultaneously to generate a single report.

7. Amassing the Data Base

A common shortcoming of existing automatic identification projects is that they have emphasized the methodology and the programming almost to the exclusion of the data collection. Methodology understandably becomes an end in itself when a new method is being developed. At this stage, however, the lack of a critical mass of data seems to be impeding further advances toward the implementation of a so-called automatic identification system. The system that wins the race will not necessarily be the one that is the most elegant in its theory and machine strategy, but rather the system that is the first to be married to an operational data base of larger than pilot proportions. By comparison with the task of gathering a critical mass of data, the task of perfecting the software seems trivial from here forward.

THE FNA GENERALIZED SYSTEM FOR PLANT DESCRIPTION

The challenge facing FNA was to design a system that could serve several taxonomic purposes equally well. For reasons already discussed, it was decided that a system designed primarily to identify plants or to construct identification keys necessarily would have to be so specialized that more generalized comparison and correlation would be precluded. On the other hand, a data base designed primarily for general description and correlation would not necessarily contain, at least during its initial years, all of the diagnostic characteristics required for identification and key construction, even if the system could be generalized to perform these functions as well as general retrieval. Identification often hinges on traits that are not likely to be described or cataloged in the course of general description but only as the result of a specific search for discriminating characteristics suitable for a dichotomous key. Key construction, as discussed earlier, hinges on the precise way in which the diagnostic characters are phrased.

Once it became apparent that a single system could not be designed to perform the three main functions of comparative description, key construction, and "automatic" identification equally well, a compromise solution was adopted. The decision was made to drop key construction altogether, as a more or less "automatic" function, from the initial concept of the system and to focus primarily on comparative description and the generalized correlation and data retrieval that it would support. As for identification, the long-term goal was to accumulate a sufficiently exhaustive data base to enable the user to discriminate most species by matching features of an unknown against the stored features of knowns. Perhaps, it was understood, the data base, by its very nature, never could provide a guaranteed delineation of every last taxon included. Thus, instead of automatic identification being the main purpose or function of the system, it would become an invaluable by-product but only *after* the comparative data base was already compiled.

1. *Design Premises and Parameters*

Consideration of the principles discussed in the foregoing sections led FNA to adopt, implicitly or explicitly, several important premises and parameters as a basis for the concept and strategy of its system. These may be stated as follows:

(a) The system is to be designed and the data base is to be created for general querying. Emphasis is to be placed, therefore, on objective comparison and correlation of descriptive data, not on diagnosis and identification *per se*. Diagnostic data are to be among the first to be incorporated into the data base, however, so that to the maximum extent possible the identification function can be served by application of the querying function, using Boolean statements to match and eliminate values.

(b) Data representation and logical structure are to be as generalized as possible in the data bank so that the individual elements can be manipulated relatively free of context and can be concatenated as building blocks into descriptive statements of any number and length as the need demands. This means that characters and character-states will be factored down to their lowest common denominators and that the terms to describe them, whether individual words, whole phrases, or numerical values, will be used so as to have a singular and explicit meaning. Every character is to be stated in full, with express indication of any multi-level hierarchy or nesting. All relationships are to be stated in the primary record or indicated by direct cross reference to other files in the data bank.

(c) Taxonomic nomenclature is to be standardized authoritatively by means of a checklist of names, which may or may not contain other information, to be known as the Taxon Name List (TNL).

(d) The descriptive vocabulary is to be controlled as to terms, classification, format, and word order. Only objectively comparative terms are to be used. In this manner, description can be restricted to the permutation of a finite number of basic, defined or definable terms that are used over and over again.

(e) Minimum requirements for data collection are to be established and observed with respect to systematic groups and geographic areas covered and characteristics described, so as to ensure some basic degree of comparability throughout the data base and thereby to minimize if not eliminate the problem of missing data. Certain characters are relatively universal, and effort is to be concentrated on building up data files for such "universals".

(f) The primary product is to be a Flora, and the data will be represented and structured in the data bank with this print requirement uppermost. To the extent that they must be predefined, the print requirements of other proposed products will be made compatible with or else subordinate to those of the Flora. Every effort is to be made to generalize the data base in the direction of keeping the printing options open, while realizing that there is no way to create a data bank in concrete form without restricting printing options, perhaps severely.

2. System Overview

An overview of the information system design for Flora North America, from the processing standpoint, has been published by Krauss (1973b), Systems Development Manager for FNA. Shetler (1975) has summarized the FNA System from the botanical standpoint, illustrating the different files with actual samples. Guidelines for contributing data to the FNA Data Bank are set forth by Shetler *et al.* (1973) and Porter *et al.* (1973). Only a capsular summary need be given here, therefore.

As explained by Krauss, the FNA System is based on the use of an information processing software known as the Generalized Information System (GIS), developed and supported by IBM and marketed through various commercial service bureaus.

Two types of files make up the FNA Data Bank: (a) authority files, and (b) data files. The chief file or library of files of data is called the Taxon Data Bank (TDB)—"Bank" as opposed to "File" in this case because it is expected to develop into a series of files.

The TDB will comprise, eventually, the species-by-species descriptions—morphological, geographical, ecological—of the approximately 20 000 species of vascular plants in North America north of Mexico. Thus it will contain 20 000 individual records, keyed by taxonomic name. Each record is likely to become quite large and will be structured hierarchically, with several to many repeating segments, as explained by Krauss (1973a, b) and Shetler (1975).

An authority file constitutes a specialized type of data file. It may be likened to a glossary or index that establishes the authoritative usage for a vocabulary of terms. The authority file, as conceived in the FNA System, constitutes the standard vocabulary for a particular category of data, e.g. taxonomic names. It controls absolutely the use of terms, as to their meaning, gender, number, spelling and form, order, and classification. A term cannot be admitted to the TDB unless it belongs to the proper authority file and is used in absolute conformity with the standards. An entry or record in an authority file may include, in addition to the term itself, such other information as is needed to define or document the term. Thus the authority file becomes a valuable though specialized data file in its own right, from which specialized reports can be published. Authority files may be seen, therefore, as comprising the elements or building materials from which the TDB is synthesized. Each descriptive record in the TDB is a unique selection and permutation of elements from the authority files, and every term in the description necessarily is derived from one of the authority files in the larger FNA Data Bank.

The following basic authority files are already in an advanced state of development:

(a) Taxon Name List (TNL)—governs taxonomic nomenclature; really two files, one for adopted names and one for synonyms.

(b) Type Specimen Register—documents typification of names in the TNL.

(c) Author File (AUF)—documents authors of taxa and defines how to use their names, whether in full or abbreviated.

(d) Morphological and Ecological Vocabulary (MEV)—determines and defines (ultimately) acceptable character and character-state terminology for describing the morphological and ecological features of the plants, and establishes the exact order and form in which the terms may be used.

(e) Specialist File—biographical documentation for taxonomic specialists contributing to the TDB or to other files in the FNA Data Bank, such as the authority files. Developed initially as a file of *potential* contributors.

Other authority files are contemplated for the vernacular names of the plants and for geographical terms. New files will be developed as the need demands.

Creation of the TDB has not begun yet, but authority-file development is now sufficiently advanced for work on the TDB to go forward.

3. Scheme for Describing Morphology

The major principles governing morphological description in the FNA System have been set out by Shetler *et al.* (1973), and the specific methods explained by Porter *et al.* (1973), in the first and second parts, respectively, of "A Guide for Contributors to Flora North America". Included in Part II, compiled by Porter *et al.*, is a provisional edition—outlined in accepted, classified order—of the exhaustive list of morphological characters (28 000) presently contained in the MEV authority file of the FNA Data Bank. Only characters in this list are to be described, but, of course, only a small fraction of the total list can be described in the average description. Of the 28 000 potential characters, the "Guidebook" specifies about 350, insofar as they apply, as constituting the minimum number required in every description. Probably no more than 25% of these are applicable on the average. The 350 characters are enumerated on the standard data sheet (Fig. 1) for describing the taxa (see Part I of "Guidebook").

The rationale for a minimum requirement of specified characters is that only by this means can true comparability be assured from one description to another. The price of making the descriptions comparative is the recording of a much

Family _____

Morphological Data Form

TDB No.
Editor

Genus _____ Species _____ Subsp./Var. _____ Contributor _____ Date

Sequence and terminology of plant parts, characters and character states should follow the Morphological Outline and Glossary in FNA Guidebook Part II. Complete one form for every taxon, regardless of rank.

DIAGNOSTIC DESCRIPTION

_____

_____

_____

_____

_____

_____

COMPARATIVE DESCRIPTION (Minimum Data Requirements)

PLANTS
duration_____
habit_____
height_____
sex_____
LEAVES
presence_____
leaf type_____
arrangement_____
BLADES
outline_____
length_____
width_____
apex shape_____
base shape_____
margin type_____
Adaxial surfaces
vestiture type_____
Abaxial surfaces
vestiture type_____
LEAFLETS
number_____
outline_____
length_____
width_____
apex shape_____
base shape_____
margin type_____
Adaxial surfaces
vestiture type_____
Abaxial surfaces
vestiture type_____
PETIOLES
presence_____
length_____
vestiture type_____
BASAL LEAVES
presence_____
leaf type_____
arrangement_____
BLADES
outline_____
length_____
width_____
apex shape_____
base shape_____
margin type_____
Adaxial surfaces
vestiture type_____
Abaxial surfaces
vestiture type_____

LEAFLETS
number_____
outline_____
length_____
width_____
apex shape_____
base shape_____
margin type_____
Adaxial surfaces
vestiture type_____
Abaxial surfaces
vestiture type_____
PETIOLES
presence_____
length_____
vestiture type_____
CAULINE LEAVES
presence_____
leaf type_____
arrangement_____
BLADES
outline_____
length_____
width_____
apex shape_____
base shape_____
margin type_____
Adaxial surfaces
vestiture type_____
Abaxial surfaces
vestiture type_____
LEAFLETS
number_____
outline_____
length_____
width_____
apex shape_____
base shape_____
margin type_____
Adaxial surfaces
vestiture type_____
Abaxial surfaces
vestiture type_____
PETIOLES
presence_____
length_____
vestiture type_____
SUBMERGED LEAVES
presence_____
leaf type_____
arrangement_____

FIG. 1. Standard FNA form for compiling morphological data on a taxon-by-taxon basis (first of four pages).

larger set of characters on the average than is typical of conventional diagnostic descriptions, so that a critical mass of comparative data, which are pertinent more or less across the plant kingdom, will be amassed. Because the same set of characters is described for all taxa, whether or not they have diagnostic value in the particular group, additional characters beyond the standard set generally must be described to ensure that the proper diagnostic features also will be included. The FNA guidelines require a fully diagnostic description in addition to the standard comparative description (see Fig. 1), except that information is not to be duplicated when there is overlap.

As explained earlier, characters (descriptors) have a nested relationship to each other. Thus, for example, the stem is a part of the plant, and the leaf in turn attaches to the stem; the rootlet is a part of the root, and the root in turn is a part of the plant, etc. This character-nesting or hierarchy may be simple, as in describing the duration or habit of the plant as a whole, but the nesting also may be complicated, breaking down into many levels from the overall plant to the ultimate part or character. The levels of nesting that one recognizes in any classified hierarchy of characters must be kept, in the final analysis, to a relatively small, arbitrary number, because, intrinsically, there is no absolute limit. A compromise must be struck between the two-level situation of describing plant habit, for example, and the multi-level situation that can arise in describing, for instance, the vestiture of a particular surface of a particular plant part.

Thus, the potentially open-ended hierarchy, for practical purposes, must be closed—telescoped into a few arbitrarily fixed levels. Consequently, the terms of any one level of headings or subdivisions will not necessarily be parallel in status. Both PLANTS and LEAVES, for example, may serve as primary headings. A purely functional point of view must be taken. These are the terms by which the data base is to be searched or sorted or the descriptions are to be organized in printed form, as in the Flora. One has come to expect such artificial subject classifications of information in conventional taxonomic descriptions, and in using the computer it is advisable to be guided by previous convention. Taxonomists have become accustomed to descriptions that treat LEAF as a heading on a par with PLANT, or STAMEN on a par with FLOWER. The taxonomist understands this kind of compromise with absolute logic to avoid the endless descriptions that would result from being totally consistent and explicit. By virtue of being predefined and fixed, however arbitrarily, the character classification, for machine purposes, is totally explicit and functions as though it were totally consistent.

Evaluation of the nesting of a representative cross-section of taxonomic characters led to the adoption of a four-level character hierarchy. This is to say that a plant may be subdivided into as many as four levels of structure in order to specify a character precisely. At least two levels must be indicated, because a character cannot be expressed without specifying the first and fourth levels, but the intermediate second and third levels are optional. Any larger hierarchy must be condensed to four levels as naturally as possible. The actual values or character-states, which describe the characters, are appended to the lowest (fourth), most specific level.

A given character may be described by any number of character-states, and the character-states may be expressed in words or numbers, including statistics. Any number of modifiers (qualifiers) may be appended to each character-state, provided that approved modifiers are used.

Three examples will serve to illustrate the principle of the four-level character hierarchy:

Example A. Cauline leaves usually whorled, sometimes opposite

Level 1.   CAULINE LEAVES
Level 2.   —
Level 3.   —
Level 4.   ARRANGEMENT   WHORLED   USUALLY
                               OPPOSITE   SOMETIMES

Example B. Ovaries 3–6

Level 1.   FLOWERS
Level 2.   GYNOECIUM
Level 3.   OVARIES
Level 4.   NUMBER 3–6

Example C. Outer surface of sepals densely lanate or tomentose

Level 1.   FLOWERS
Level 2.   SEPALS
Level 3.   ABAXIAL SURFACES
Level 4.   VESTITURE TYPE   LANATE   DENSELY
                                TOMENTOSE   DENSELY

If these character descriptions were to be printed out by computer exactly as stored with the full hierarchy explicitly stated, they would read quite unnaturally:

(A) CAULINE LEAVES ARRANGEMENT WHORLED USUALLY OPPOSITE SOMETIMES
(B) FLOWERS GYNOECIUM OVARIES NUMBER 3–6
(C) FLOWERS SEPALS ABAXIAL SURFACES VESTITURE TYPE LANATE DENSELY TOMENTOSE DENSELY

In fact, the scheme was devised on the assumption that only selected words would be printed. This explains why certain words are plural, others singular. Selective printing was to be accomplished by including a print or not-print code with each level in every character description as stored in the computer.

With selective printing, these character descriptions would read almost like conventional descriptions:

(A) CAULINE LEAVES WHORLED USUALLY OPPOSITE SOME-TIMES

(B) OVARIES 3–6

(C) (SEPALS—understood by context) ABAXIAL SURFACES LANATE DENSELY TOMENTOSE DENSELY

Finally, it was expected that the report-generating program would be refined still further until word order could be reversed and necessary punctuation and connectives could be added as desired. Modifiers then could be placed in their conventional position, and the character-states could be separated by the appropriate punctuation marks ("cauline leaves usually whorled, sometimes opposite" instead of "cauline leaves whorled usually opposite sometimes").

A concrete example of a species description is illustrated in tabular form in Table I. It is a partial description of *Xylosma flexuosa* (H.B.K.) Hemsl. (Flacourtiaceae), the "Expanded Description" in Porter *et al.* (1973, p. viii). The tabular organization approximates the logical structure of a description in the data bank but should not be construed as the physical representation in magnetic storage.

In Table I, the full hierarchy is spelled out for each character described, as it must be recorded in the data bank. If necessary, the hierarchical classification can be compressed, reducing the indentation and spacing to a minimum. In Part II of the "Guidebook", where the terms are all printed by computer in capital letters, typography is uniform and, therefore, of no help in differentiating the four levels of hierarchy. Consequently, the following system of compressed and numbered indentation was used, and it conserved enough space to print three columns of the classified vocabulary on each page:

1 2 3 4
BRANCHES
 PRICKLES
  SURFACE
   COLOR
   VESTITURE TYPE
   ORIENTATION
[ETC.]

Here the numbers at the top specify the columns in which the first, second, third, and fourth levels of the indented hierarchy begin.

On the data form (Fig. 1), the hierarchical classification of characters is represented by varying the size and case of the letters in combination with compressed spacing and indentation. Blanks are provided for the actual character states, including modifiers. For the user's sake, the hierarchical classification of plant parts and characters was telescoped and simplified on the Data Form as compared with the Guidebook, but without changing the relative order of relationships.

Table I provides a vivid example of the order of magnitude of expansion when a typical, telegraphic, taxonomic description is blown up into data bank format. Everything must be made explicit. Thus, while to the botanist duration is implicit in the term "shrub", the computer must be "told" that the plant is a perennial. Otherwise, any later machine search for perennials will pass by this species. Likewise, it is necessary to state explicitly that such plant parts as branches, twigs, and leaves are "present".

The conventional version of this description, as it would be printed in a Flora or monograph, consists of the italicized terms in the table (reading from left to right and top to bottom) and would print out approximately as follows, if one assumes that the report-generating program would correct word order and insert essential punctuation:

SHRUBS TO 2 M HIGH, POLYGAMODIOECIOUS; WOOD
MODERATELY HARD, DIFFUSE-POROUS; BRANCHES SLENDER
OR STOUT, GLABRATE; TWIGS DENSELY PUBERULENT OR
GLABRESCENT; SPURS OFTEN PRESENT; THORNS OFTEN
PRESENT, TO 33 MM LONG, SLENDER OR STOUT; LEAVES
REMOTE OR SOMETIMES CROWDED (ON SPURS); BLADES
RHOMBIC-OVATE, OVATE-ELLIPTIC, ELLIPTIC,
OBLANCEOLATE OR OBOVATE, CHARTACEOUS OR
SUBCORIACEOUS, TO 6 CM LONG, TO 3 CM WIDE; APEX
ACUMINATE, ACUTE, OBTUSE OR ROUNDED; BASE
ATTENUATE-CUNEATE OR CUNEATE; MARGIN IRREGULARLY
CRENULATE-SERRULATE; ADAXIAL SURFACES GREEN, OFTEN
LUSTROUS, GLABROUS; PRIMARY VEINS PUBERULENT:
SECONDARY VEINS 5–6 PER SIDE, EVIDENT; TERTIARY VEINS
INCONSPICUOUS; ABAXIAL SURFACES PALER GREEN,
GLABROUS; SECONDARY VEINS EVIDENT; TERTIARY VEINS
INCONSPICUOUS; PETIOLES SHORT; INFLORESCENCES
FASCICULATE, SESSILE; PEDICELS TO 5 MM LONG, GLABRATE

H

Table I. A partial description of *Xylosma flexuosa* (H.B.K.) Hemsl. (Flacourtiaceae) in FNA Data Bank format (see "Guidebook", Part II, Potter *et al.*, 1973)

Primary Part	Secondary Part	Tertiary Part	Character	Character-State	Modifier	Guidebook Reference
PLANTS	—	—	DURATION	PERENNIAL	—	p. 1
PLANTS	—	—	HABIT	SHRUBS*¹	—	p. 1
PLANTS	—	—	HEIGHT	2 M[HIGH]*²	TO*³[UP TO]	p. 1
PLANTS	—	—	SEX	POLYGAMODIOECIOUS*	—	p. 1
MAIN STEMS	WOOD	—	TEXTURE	HARD	MODERATELY³	p. 1
MAIN STEMS	WOOD	—	WOOD POROSITY	DIFFUSE-POROUS	—	p. 1
BRANCHES	—	—	PRESENCE	PRESENT	—	p. 2
BRANCHES	NODES	—	DIAMETER	SLENDER STOUT	[OR]	p. 2
BRANCHES	NODES	VESTITURE	VESTITURE TYPE	GLABRATE	—	p. 2
BRANCHES	INTERNODES	—	DIAMETER	SLENDER STOUT	[OR]	p. 2
BRANCHES	INTERNODES	VESTITURE	VESTITURE TYPE	GLABRATE	—	p. 2
TWIGS	—	—	PRESENCE	PRESENT	—	p. 3
TWIGS	NODES	VESTITURE	VESTITURE TYPE	PUBERULENT GLABRESCENT	DENSELY³[OR]	p. 3
TWIGS	INTERNODES	VESTITURE	VESTITURE TYPE	PUBERULENT GLABRESCENT	DENSELY[OR]	p. 3
SPURS	—	—	PRESENCE	PRESENT	OFTEN³	p. 10
THORNS*	—	—	PRESENCE	PRESENT*	OFTEN*³	p. 12
THORNS	—	—	LENGTH	33 MM[LONG]	TO³[UP TO]	p. 12
THORNS	NODES	—	DIAMETER	SLENDER STOUT	[OR]	p. 12

THORNS	INTERNODES	—	DIAMETER	SLENDER	[OR]	p. 12
				STOUT		
LEAVES	—	—	PRESENCE	PRESENT	—	p. 14
LEAVES	—	—	PROXIMITY	REMOTE	[OR]	p. 14
LEAVES	—	—	ARRANGEMENT	CROWDED (ON SPURS)	SOMETIMES[3]	p. 14
BLADES	—	—	OUTLINE	RHOMBIC-OVATE*	[.]*	p. 14
				OVATE-ELLIPTIC*	[.]	p. 14
				ELLIPTIC*	[.]	p. 14
				OBLANCEOLATE*	[OR]	p. 14
				OBOVATE*	—	p. 14
BLADES	—	—	TEXTURE	CHARTACEOUS*	[OR]*	p. 14
				SUBCORIACEOUS*		
BLADES	—	—	LENGTH	6 CM[LONG]*	TO*[3][UP TO]	p. 14
BLADES	—	—	WIDTH	3 CM[WIDE]*	TO*[3][UP TO]	p. 14
BLADES	APEX*	—	APEX SHAPE	ACUMINATE*	[.]*	p. 14
				ACUTE*	[.]*	
				OBTUSE*	[OR]*	
				ROUNDED*	—	
BLADES	BASE*	—	BASE SHAPE	ATTENUATE-CUNEATE*	[OR]*	p. 14
				CUNEATE		
BLADES	MARGIN*	—	MARGIN TYPE	CRENULATE-SERRULATE*	IRREGULARLY*[3]	p. 14
BLADES	ADAXIAL SURFACES	SURFACE	COLOR	GREEN	—	p. 15
BLADES	ADAXIAL SURFACES	SURFACE	SURFACE CHARACTER	LUSTROUS	OFTEN[3]	p. 15
BLADES	ADAXIAL SURFACES	VESTITURE	VESTITURE TYPE	GLABROUS	—	p. 15
BLADES	ADAXIAL SURFACES	PRIMARY VEINS	VESTITURE TYPE	PUBERULENT	—	p. 15

Table I.—continued

Primary Part	Secondary Part	Tertiary Part	Character	Character-State	Modifier	Guidebook Reference
BLADES	ADAXIAL SURFACES	SECONDARY VEINS	GENERAL	EVIDENT	—	p. 15
BLADES	ADAXIAL SURFACES	SECONDARY VEINS	NUMBER PER SIDE	5–6[PER SIDE]	—	p. 15
BLADES	ADAXIAL SURFACES	TERTIARY VEINS	GENERAL	INCONSPICUOUS	—	p. 15
BLADES	*ABAXIAL SURFACES*	SURFACE	COLOR	GREEN	*PALER*[3]	p. 15
BLADES	ABAXIAL SURFACES	VESTITURE	VESTITURE TYPE	GLABROUS	—	p. 15
BLADES	ABAXIAL SURFACES	SECONDARY VEINS	GENERAL	EVIDENT	—	p. 15
BLADES	ABAXIAL SURFACES	TERTIARY VEINS	GENERAL	INCONSPICUOUS	—	p. 15
*PETIOLES**	—	—	PRESENCE	PRESENT	—	p. 24
PETIOLES	—	—	LENGTH	*SHORT**	—	p. 24
*INFLORESCENCES**	—	—	INFL ARRANGEMENT	FASCICULATE*	—	p. 29
INFLORESCENCES	—	—	POSITION	SESSILE*	—	p. 29
*PEDICELS**	—	—	LENGTH	5 MM[LONG]	*TO**[UP TO]	p. 32
PEDICELS	—	VESTITURE	VESTITURE TYPE	GLABRATE	[OR]	p. 32
				PUBERULENT		
*SEPALS**	—	—	PRESENCE	PRESENT	—	p. 65
SEPALS	—	—	NUMBER	5*	—	p. 65

Organ	Part	Character	State	Modifier / Connective	Page
SEPALS*	—	OUTLINE	OVATE-OBLONG	—	p. 65
SEPALS	—		1-2 MM[LONG]*	[OR]	p. 65
SEPALS	APEX	APEX SHAPE	ACUTE	—	p. 65
			OBTUSE		p. 65
SEPALS	MARGIN*	MARGIN TYPE	CILIATE*	CONSPICUOUSLY*	p. 65
SEPALS	ABAXIAL SURFACES	VESTITURE	VESTITURE TYPE	GLABROUS	[OR]	p. 65
			PUBERULENT		p. 65
FILAMENTS*	—	LENGTH	3-4 MM[LONG]*	—	p. 68
GYNOECIUM	OVARIES	CARPELS	NUMBER	2-6	—	p. 71
GYNOECIUM	OVULES	OVULE NUMBER PER LOCULE	2-8	—	p. 71
GYNOECIUM	STYLES	DURATION	PERSISTENT	—	p. 71
GYNOECIUM	STYLE BRANCHES*	NUMBER	2*	—	p. 71
GYNOECIUM	STIGMAS	NUMBER	2	—	p. 71
GYNOECIUM	STIGMAS	SHAPE	DILATED	—	p. 71
FRUITS*	—	SHAPE	SUBGLOBOSE*	—	p. 118
FRUITS	—	DIAMETER	6 MM[DIAMETER]*	TO*[UP TO]	p. 118
FRUITS	—	COLOR	RED*	—	p. 118
FRUITS	SURFACE	VESTITURE	VESTITURE TYPE	GLABROUS*	—	p. 118
SEEDS*	—	LENGTH	3-4 MM[LONG]	—	p. 119
SEEDS	—	DIAMETER	2-3 MM[DIAMETER]	—	p. 119
SEEDS	—	SEED NUMBER PER FRUIT	2-8*	—	p. 119

[1] The terms, modifiers, and connectives (including punctuation) printed in italics constitute the description as it would be presented in written form (reading from left to right and from top to bottom). These are the words and marks that would be coded for print-out. The diagnostic description consists of the subset of these terms marked by asterisks (*).

[2] Interpolated words are placed in brackets [ ] and italicized where essential to the sense of the descriptive statement.

[3] These modifiers should be placed before rather than after the character-state when printed.

OR PUBERULENT; SEPALS 5, OVATE-OBLONG, 1–2 MM LONG; APEX ACUTE OR OBTUSE; MARGIN CONSPICUOUSLY CILIATE; ABAXIAL SURFACES GLABROUS OR PUBERULENT; FILAMENTS 3–4 MM LONG; CARPELS 2–6; OVULE NUMBER PER LOCULE 2–8; STYLES PERSISTENT; STYLE BRANCHES 2; STIGMAS 2, DILATED; FRUITS SUBGLOBOSE, TO 6 MM DIAMETER, RED, GLABROUS; SEEDS 3–4 MM LONG, 2–3 MM DIAMETER; SEED NUMBER PER FRUIT 2–8.

The diagnostic features embedded in this description are the ones starred (*) in Table I and of course would compile into a shorter paragraph.

This data format imposes four main requirements on the computer-processing strategy. First, all relationships of the character classification must be maintained, which is to say that both the hierarchy (nesting) and the ordering (arrangement) of the characters must be held constant throughout all taxonomic descriptions in the data bank. Second, every character must constitute a unique combination of terms in levels 1–4 (i.e. primary part, secondary part, tertiary part, character), so that in and of itself it describes one and only one feature of the plant. Third, there must be a way to repeat character-states or modifiers without repeating the character or character-state, respectively. This is especially important if variations in the statement of a character or a character-state are to be avoided when multiple values are appended. Fourth, the processing strategy must provide a means of compressing the data. Otherwise, the size of the data base would quickly get out of hand as the bulky descriptions, such as the one illustrated in Table I, were compiled for upwards of 20 000 species, and the cost of processing would become astronomical.

The solution adopted by FNA is to generate a parallel numerical classification or *numericlature* and then to store and process the character statements as numerical codes using software capable of supporting hierarchical data structures. Only the actual character-states and their modifiers are stored as words in the individual records. The numerical codes constitute the cross links to the Morphological and Ecological Vocabulary (MEV), the authority file governing the classification and order of the characters, where the numericlature is equated to its parallel nomenclature (see earlier discussion about the MEV).

The numericlature is generated by assigning a unique, systematic, numeric code of fixed length to every unique term in each of the four levels of the character hierarchy, i.e. primary part, secondary part, tertiary part, and character (e.g. BRANCHES, INTERNODES, VESTITURE, VESTITURE TYPE). For each level or column (see Table I), the vocabulary of terms is established in some absolutely fixed order, and then the terms are numbered consecutively from beginning to end. The numbering is done by tens or hundreds to allow for the interpolation of new terms in the future as needed without changing the relative order of the original terms. Thus, assuming for the sake of illustration that the terms on the first page of the data form shown in Fig. 1 represent a closed vocabulary and that the numbering is done by tens, we have five terms in Column (Level) 1 that code as follows:

PLANTS	10
LEAVES	20
BASAL LEAVES	30
CAULINE LEAVES	40
SUBMERGED LEAVES	50

Any term interpolated between PLANTS and LEAVES will have a number between 10 and 20, say 15, etc.

In theory, at least, the original numbers are not changed once their assignment is definitive, i.e. after all editorial vetting and debugging, regardless of how many additions or deletions there are. If terms are deleted, their equivalent codes are deleted, and the remaining terms and codes retain their original relative order. This can only work if in the original numbering rather large gaps are left in the codes between successive terms, to allow for all possible future additions, and if the final order can be established at once. In practice, the ordering becomes the most difficult problem, and it may be necessary to regenerate the numericlature several times over a period of time before the order is satisfactory. Until data banking based on the numericlature has begun, regenerating the codes does not pose an extraordinary problem. Once data banking has begun, however, then making a change in the code of a particular term in the authority file (i.e. the MEV) necessitates changing the code wherever this term has been used in other files in the data bank. The impact of such changes is at once apparent. It multiplies rapidly as the files grow.

Returning to our example in Figure 1, there are three unique terms in Column (Level) 2, which code as follows:

BLADES	10
LEAFLETS	20
PETIOLES	30

The two unique terms in Column 3 and their codes are:

ADAXIAL SURFACES	10
ABAXIAL SURFACES	20

The fourth level, the Character Level, presents the greatest difficulty. The number of unique terms is much larger, although more limited than might be supposed, but the ordering presents the greatest difficulty. To illustrate without arguing order, let us take the order given in Fig. 1:

DURATION	010
HABIT	020
HEIGHT	030
SEX	040
PRESENCE	050
LEAF TYPE	060
ARRANGEMENT	070
NUMBER	080
OUTLINE	090
LENGTH	100
WIDTH	110
APEX SHAPE	120
BASE SHAPE	130
MARGIN TYPE	140
VESTITURE TYPE	150

The full numericlature of a character statement consists of the set of codes for each of the four levels of hierarchy taken together—the numbers for a *row* of terms as presented in Table I. Examples based on the above codes generated for Fig. 1 are:

10 00 00 010   (DURATION of PLANTS)
20 10 00 090   (OUTLINE of BLADES of LEAVES)
40 10 00 090   (OUTLINE of BLADES of CAULINE LEAVES)
40 20 20 150   (VESTITURE TYPE of ABAXIAL SURFACES of
                CAULINE LEAVES)
40 30 00 100   (LENGTH of PETIOLES of CAULINE LEAVES)

Given this approach to the coding, which assigns a unique number from a consecutive sequence to each unique term at each level, the numericlature guarantees that the unique terms of a particular column (level) will always sort numerically into the same order. Likewise, it guarantees that a single order will always result when the terms are sorted *hierarchically* by number, i.e. second-level categories (terms) within first-level categories, third-level within second-level categories, and fourth-level within third-level. No matter how many times a term (category), e.g. OUTLINE, repeats itself among categories of a higher level, it will always sort into the same order relative to other terms. To illustrate, the terms NUMBER, OUTLINE, and MARGIN TYPE of the fourth-level (character-level) vocabulary may be used repeatedly to describe different parts of the plant as delineated in the categories of the third or higher levels, but always

NUMBER will precede OUTLINE and both will precede MARGIN TYPE *within category*. Furthermore, the relative order of terms in any subset, regardless how large or small, will be identical to that of the entire vocabulary, whether the columns are taken individually or collectively. Finally, within level a particular numerical code can have only one nomenclatural equivalent, whereas across two or more levels a particular combination of codes, i.e. a particular numericlatural value, can point to only one combination of terms. Thus, the codes can serve in place of the words for the purposes of search and comparison.

In the data bank, the characters of a plant are designated by their appropriate numerical codes, and the actual values or character-states and their modifiers are appended, as words either in full or abbreviated. Some examples based on Fig. 1 and the above codes might be:

```
10 00 00 010   PERENNIAL
               USUALLY
               BIENNIAL
               SOMETIMES
  (DURATION of PLANTS USUALLY PERENNIAL, SOMETIMES
     BIENNIAL)
20 10 00 090   OVATE
                 or
               LANCEOLATE
               BROADLY
  (OUTLINE of BLADES of LEAVES OVATE or BROADLY
     LANCEOLATE)
40 20 20 150   PUBERULENT
               FINELY
  (VESTITURE TYPE of ABAXIAL SURFACES of CAULINE
     LEAVES FINELY PUBERULENT)
```

The numerical codes are further augmented in the computer record by coded printing instructions, as needed. To the code in each column must be added one more digit or a letter to indicate whether the term represented by the code (e.g. 10=PLANTS) should be printed automatically by machine when a description is composed. Further coding may be necessary in the fourth column or following the character-state or its modifier to indicate what type of punctuation or connectives are needed and whether the modifier should precede or follow the character-state. Thus, in the example above concerning plant duration, the machine must be able to transform 10 00 00 010 PERENNIAL USUALLY BIENNIAL SOMETIMES into PLANTS USUALLY PEREN-NIAL, SOMETIMES BIENNIAL. This involves printing the first-level term,

PLANTS, and the character-states and their modifiers properly arranged an
punctuated, but not printing the second-, third-, and fourth-level terms, whic
in the first two instances are blanks but in the third case would be the wor
DURATION. The print codes will not be constant for any particular terms bu
presumably will be constant for particular *combinations* of terms. How th
print procedure works can be appreciated by studying the *italicized* words i
Table I.

The FNA character classification (Porter *et al.*, 1973) consists of approximatel
28 000 lines or rows of unique combinations of first-, second-, third-, and fourth
level terms. In order to allow for all of the potentially numerous new terms tha
might be added to each level, the original terms were numbered consecutivel
by hundreds, necessitating a five-digit code at each level. Thus the basi
numericlature consists of 20 digits, and following are some examples:

SECOND GLUME	AWNS	—	LENGTH
			04770 02350 00000 07700
SECOND GLUMES	AWNS	TRICHOMES	LENGTH
			04770 02350 01400 07700
CALYX	—	—	POSITION
			05004 00000 00000 05000
CALYX	LOBES	MARGIN	
		TRICHOMES	LENGTH
			05004 01800 04000 07700
CALYX	APPENDAGES	MARGIN	
		TRICHOMES	SHAPE
			05004 02900 04000 09100

It should be added that these codes are intended for internal use in the data
banking operation. The outside contributor or user need not concern himsel
with the codes, and for this reason they are not printed in Part II of th
"Guidebook" along with the classified vocabulary.

Figure 2 is a schematic representation of the logical data hierarchy (inverte
"branching tree") from the point of view of the processing system, GIS. Th
hierarchical file structures supported by GIS are explained in detail by Kraus
(1973a, b). It should be emphasized that the data hierarchy in this instance is no
the same as the character hierarchy or nesting, nor does it have anything to d
with taxonomic hierarchy.

This is a three-level hierarchy. The Fixed or Root Segment specifies the plan
character being described, according to the four-level classification of character
explained at length above. In the authority file, i.e. the MEV, the terms and th
equivalent numericlature are given; in the Taxon Data Bank only th

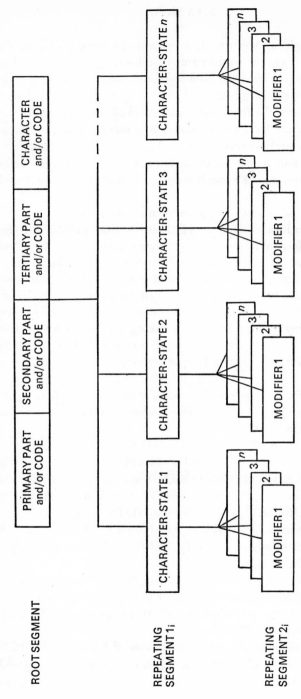

Fig. 2. Schematic representation of the logical, three-level, data hierarchy for a morphological description in the Taxon Data Bank (TDB) as structured for processing by the Generalized Information System (GIS). The Root Segment alone is used in the backup authority file, the Morphological and Ecological Vocabulary (MEV), but with both words and codes, whereas all levels are used in the TDB, but only codes are entered in the Root Segment.

numericlature is given. Segment 1_i represents the second level of the dat. hierarchy. This segment carries the term(s) for describing a particular character state. It may repeat relative to the Root Segment as often as necessary to describ all the character-states (Segment $1_1, 1_2, 1_3, \ldots 1_n$) of the specified character Segment 2_i represents the third and lowest level of the data hierarchy whicl carries the modifiers. It may repeat relative to a particular character-state (e.g Segment 1_1) as often as necessary (Segment $2_1, 2_2, 2_3, \ldots 2_n$) to modify the given character-state. Herein lies the branching tree: one character may be defined by many character-states each of which in turn may be qualified by many modifiers

USE OF FNA DATA BANK TO IDENTIFY PLANTS

The possibilities for identification need little elaboration at this point. The data bank content and format, illustrated by concrete example in Fig. 1, will suppor many different types of searches and comparisons to match the features of ar unknown plant against the stored descriptions of known plants. Searches can be made for all taxa answering either to a particular condition or to an elaborate se of conditions. Searches can be conducted within level (column) irrespective o the character-nesting relationships or by taking into account the character-defining hierarchy of plant parts. Likewise character-states (e.g. PERENNIAL) can be searched with or without taking into account the parts of the plants to which they apply.

Following are a few examples of the types of searches that might be made:

(a) Find (and list) all the different character-states recorded for HABIT of PLANTS and tally the number of each.

(b) Find all POLYGAMODIOECIOUS PLANTS.

(c) Find all POLYGAMODIOECIOUS PLANTS of HEIGHT less than or equal to 2 M, which have DIFFUSE-POROUS WOOD in the MAIN STEMS.

(d) Find all instances in which the INTERNODES are GLABROUS but the NODES are PUBERULENT or otherwise NOT GLABROUS.

(e) Find all instances in which the NODES of the TWIGS are PUBERULENT or otherwise NOT GLABROUS, but the NODES of the BRANCHES are GLABROUS.

(f) Find all SHRUBS of HEIGHT less than or equal to 2 M, which have both SPURS and THORNS (or, which have either SPURS or THORNS).

(g) Find all cases of BLADES that are OBOVATE, CHARTACEOUS, and not exceeding 6 CM in LENGTH.

(h) Find all the different character-states recorded for VESTITURE TYPE and tally the number of each. Repeat the search restricting it to the ABAXIAL SURFACES of BLADES.

(i) Find coincidences of SPURS or THORNS and FASCICULATE, SESSILE INFLORESCENCES with GLABRATE or PUBERULENT PEDICELS.

(j) Find cases of GLOBOSE or SUBGLOBOSE, RED FRUITS with a SEED NUMBER PER FRUIT of 2 or more.

Obviously, the possibilities are endless if the data are present in the first place and the user knows his data base and is clever in framing Boolean statements to query it. In identification, the aim is to find a precise and unique match with the character-state or set of character-states of the unknown specimen, but any narrowing down of the possibilities is a help. The matching process can be repeated until only one to several possibilities remain. Using a single character-state seldom will narrow the range appreciably, except when the character-state is highly unusual, but this may be the only alternative when only a scrap of a plant is available—a situation in which the conventional dichotomous key generally is useless. On the other hand, the more character-states required to be coincident, the more likely that no "hit" will be found in the data base. This problem can be surmounted by using some type of probability algorithm that requires the record to meet only some of the stipulated conditions, say 75% of them, in order to be selected (i.e. to be a "hit") during the search, and that then ranks those selected according to the degree to which they meet the conditions. The taxonomist can check out the various possibilities, perhaps by comparing his unknown with herbarium specimens of the possible entities, and decide which one, if any, to accept. In the final analysis, this approach to identification may yield more accurate results and greater satisfaction with the identifications than the traditional approach of choosing between two idealized alternatives in a dichotomous key. The traditional dichotomous key has dominated the taxonomist's approach to identification for so long that it is difficult for him to think in any other vein and particularly to visualize in the abstract what appear to be the tremendous possibilities of a generalized data-banking approach to identification.

CONCLUSION

Taxonomists always dream of the Utopian day when every living organism on earth will have been discovered and cataloged and every conceivable feature of every species will have been measured or described to the satisfaction of all earthly authority. The computer has added a new dimension to our dream. Now we are fond of imagining that one day all of these facts, once gathered, will be stored in some omniscient memory manipulated by an omnipotent machine that can think for the taxonomist, match wits with nature, and always provide the right answers—precisely, instantly, and automatically. Bounding from one hyperbole to another, we fantasize about conversing as humans with this machine, commanding it to consult its limitless memory and then to cause its system to create, package, and issue—the jargon is "print out"—dichotomous keys, descriptions, and other of the forms of organized knowledge that have been the hallmark of taxonomists through history, but with a touch of superhuman perfection in content and style, not to mention speed and efficiency.

In reality, the computer can do little to alleviate the taxonomist's historical burdens. The notion of automating the taxonomist's chores is, at this stage, I submit, still largely a figment of our rhetoric, born of experimental but not operational experience. Perhaps we now have the potential for certain types of automation, but little has been realized. No matter what the computer does for the taxonomist, however, he still will find himself stuck with the ancient and painstaking tasks of gathering, evaluating, organizing, and editing the data. He still will have to render all of the judgments (Shetler, 1974).

The much-hoped-for inexhaustible data bank surely is an impossible dream. We now realize that the exhaustive and definitive cataloging of the earth's biota is not only unrealizeable but unimaginable (Raven *et al.*, 1971). For all practical purposes, the universe of organisms is infinite. Despite the efforts of generations of specialists since Linnaeus' time to catalog this universe, the gap does not seem to close but to widen. The goal, like that elusive last mountain peak, seems to retreat as we climb toward it. Today's taxonomists probably perceive themselves as being farther from the goal, relatively speaking, than did Linnaeus and the other forerunners of his time.

It is in the nature of taxonomy that there will always be divergent philosophies and species concepts, and history has proven that each generation will have its own varieties of taxonomic "lumpers" and "splitters", who, in addition to discovering and describing what obviously is new, will continue endlessly to subdivide and recombine what is known. Add to their perennial differences of philosophy the taxonomists' perennial problems of keeping abreast of what already has been described, and we have the reason for the enormous synonymy that exists in our literature today and also the basis for predicting confidently that the process never will change. Each generation will find grounds for describing new species and discarding old ones. Thus it seems most unlikely that biology ever will be able to provide an exact and final reckoning of the kinds of organisms in the world now or in the future, no matter how much specialized manpower is devoted to this task. Indeed, differences of opinion and failures in communication will increase in direct proportion to the number of practitioners in the field and the amount of data accumulated, and one might predict safely that the possibility of achieving a total inventory of the earth's biota will become more, instead of less, remote with time.

Looking at the other factor in the data-banking equation—the number of characters to be described—we face equally bleak prospects for conclusive treatment. The universe of observable and measurable characteristics is even more obviously beyond total enumeration than the universe of taxa. Any taxonomist knows how subjective and how relative is the choosing of characteristics to describe and, therefore, how utterly impractical it is to suppose that one ever could prepare, even for a single species, an exhaustive description that would stand as complete and definitive for all time.

What conclusions can we draw, therefore? If the exhaustive and definitive data base is an impossible dream, why, then, embark on the course of computerized data banking, particularly generalized data banking? The very concept of generalization, as set forth in this paper, tends to foster the impossible dream.

In the first place, we need to recognize the basic futility of trying to catalog the earth's biota exhaustively and definitively and to begin focusing on ways to extract more information from the data that we already have in hand.

In the second place, we need to recognize that the computer can be a powerful tool in the hands of the taxonomist precisely because it can help him to permute and to manipulate his *existing* data and thus to extract more information from the same data, but not because it can help him in the original task of describing the organisms and compiling the data. The computer does not change the basic human variables in the taxonomic equation; it does not help the taxonomist to cover more ground—to describe either more organisms or more characteristics. This is to say that the computer as such does nothing to facilitate the cataloging of the earth's biota nor to increase the chances of achieving the ultimate survey. (Likewise, one might inject, in the area of systematics collections the computer of itself can do nothing to facilitate the cataloging of specimens nor to increase the chances of achieving a total inventory.) When this fundamental principle is understood, efforts to use the computer as an information retrieval tool in taxonomy will not lead to fruitless or disastrous ends.

If the computer provides the taxonomist with unparalleled new opportunities in manipulating stored data, it also brings him to an historical crossroads in plant description. The opportunities can be realized only by departing from the traditional course of purely diagnostic description and taking up the course of generalized, comparative description *but within predefined finite limits*. The Flora North America is predicated on the assumption that the time is ripe for plant taxonomists to begin taking the latter course. At least in north temperate regions, the discovery phase has reached the point of diminishing returns, and our knowledge of the flora is extensive. In the community of taxonomists, the shift from primary data-gathering and description to synthetic description and correlation has been underway for many years, and the experience of the FNA Program has shown that there is a general willingness to cooperate on a large scale in the creation of a data bank of truly comparative descriptions. The modern computer now makes it possible to take full advantage of such a data bank, provided that taxonomists can work in concert to harness the machine for this purpose.

In the short term, the benefits of a computerized taxonomic data bank may be limited, because it takes so long to build up a critical mass of data, but in the long term the benefits can be enormous just in the practical realm of plant diagnosis and identification alone, because broader and broader correlations become possible as the data base grows larger and larger and approaches the established limits. Fictions aside, the potentialities of taxonomic data banking are extra-ordinary and well worth pursuit insofar as wise and attainable goals can be set.

ACKNOWLEDGMENTS
Special thanks are due to the Program Council of Flora North America, particularly to its chairman, Peter H. Raven, and members Duncan M. Porter, John H. Thomas, and Harriet M. Krauss, and to Robert W. Kiger, Thomas E. Kopfler, Judith E. Monahan, Larry E. Morse, and Robert W. Read, for the critical roles that they played in developing the FNA concept of data banking; and to Joanne V. Wescott, for indispensable technical assistance in the preparation of this manuscript.

REFERENCES
Bailey, L. H. (1949). "Manual of Cultivated Plants" (Revised edition). Macmillan, New York.

Krauss, H. M. (1973a). The use of generalized information processing systems in the biological sciences. *Taxon* 22, 3–18. (Flora North America Report No. 67.)

Krauss, H. M. (1973b). The information system design for the Flora North America Program. *Brittonia* 25, 119–134. (Flora North America Report No. 69.)

Morse, L. E. (1974). Computer programs for specimen identification, key construction, and description printing using taxonomic data matrices. Biological series, Publications of the Museum of Michigan State University 5, 1–128, East Lansing, Michigan.

Porter, D. M., Kiger, R. W. and Monahan, J. E. (1973). A guide for contributors to Flora North America, Part II: An outline and glossary of terms for morphological and habitat description (provisional edition). Flora North America Report No. 66, p. i–x, 1–120, G1–G32.

Raven, P. H., Berlin, B. and Breedlove, D. E. (1971). The origins of taxonomy. *Science* 17, 1210–1213.

SHETLER, S. G. (1974). Demythologizing biological data banking. *Taxon* **23**, 74–104. (Flora North America Report No. 75.)

SHETLER, S. G. (1975). The Flora North America system. (Abstract) Papers of meeting on use of EDP in the herbarium, sponsored by NATO Ecosciences Panel, October 1973, Royal Botanic Gardens, Kew (in press). Flora North America Report No. 77.

SHETLER, S. G. *et al.* (1973). A guide for contributors to Flora North America (FNA) (provisional edition). Flora North America Report No. 65, p. i–ix, 1–28 + appendices A–E.

14 | Identification Methods and the Quality of Taxonomic Descriptions

R. J. PANKHURST

Department of Botany, University of Cambridge, England

Abstract: The nature of conventional taxonomic descriptions is considered, and the inadequacies of such data for use in automatic identification techniques are discussed. Statistics are quoted concerning the quality of data in one particular monograph covering a large number of species, and the problems and techniques for improving these data are described. Criteria for establishing better taxonomic descriptions are proposed.

Key Words and Phrases: identification, taxonomic descriptions, comparability of characters, completeness of taxonomic descriptions, errors in taxonomic descriptions, Floras, diagnostic keys, Rubus

NATURE OF TAXONOMIC DESCRIPTIONS

Descriptions of biological taxa are often published as written text in handbooks, manuals, monographs and floras. These descriptions serve to establish classifications, to support nomenclature or as references for identification, or some combination of these purposes. Each description refers to one specimen, or is a combined description based on several specimens of other taxa of lower rank. The material or other sources on which a description is based is not always noted. Other authors have criticized the quality of taxonomic data (cf. Heywood, 1973; Watson, L., 1971; Shetler, this volume), but here the effect on identification by computer is discussed.

The information in such descriptions is usually incomplete, that is to say, the author has not attempted to describe every character for every taxon. In some cases it is, of course, logically impossible to observe a character. For example, a leaf might, or might not, have a toothed margin. For a leaf without teeth, it would be impossible to describe the shape of the teeth. Then "shape of teeth on leaf" is a conditional character depending on presence of teeth. In many cases,

Systematics Association Special Volume No. 7, "Biological Identification with Computers", edited by R. J. Pankhurst, 1975, pp. 237–247. Academic Press, London and New York.

however, the omission has no such logical basis. For example, in an account of
the plant genus *Epilobium* (Clapham, Tutin and Warburg, 1962), the stigma is
described as either entire or four-lobed. This character is given for the first nine
species, but omitted for the last three. However, in the *Epilobium* example, all
species have stigmas. There is also a general tendency to omit median values of
characters. For example, in a monograph of the genus *Rubus* (Watson, W. C. R.,
1958) the width of petals is given as narrow or broad. Independent observations
showed that most petals could be described as "moderately broad", but this was
nearly always omitted. There is also perhaps a tendency not to describe the
extent of variation where characters are variable. A discussion of some reasons
why information is omitted is given later.

The descriptive terms used can be used inconsistently at times, that is to
say that a character of one taxon is not strictly comparable with the same
character of another. For an example, in a monograph of *Taraxacum* flower
colours are described, among other things, as "pale yellow", "(mid-)
yellow", "dark yellow", and "bright yellow". Of these "bright yellow" is
inconsistent, because it describes, presumably, the reflectivity of the "petal"
surface, rather than a colour alone. A species described as "dark yellow" cannot
be compared with another which is "bright yellow", because it is not clear that
the author is referring to the same character. A test for this situation is to ask
whether the character states are exclusive or not. A consistent way of using a
character is to use states which are mutually exclusive, e.g. "dark yellow" and
"pale yellow" cannot be true simultaneously, and they are mutually exclusive,
and therefore consistent. However, a petal could be both "mid yellow" and
"bright yellow", so these are not exclusive and not consistent. Inconsistency in
the use of characters is much more marked when works by different authors are
compared, and can only be resolved if each author carefully defines the terms
he uses.

A diagnostic key is often provided together with the descriptive text, and
some authors make a practice of including factual information in the key which
is omitted from the text. Not all authors ensure that the key and the text agree
in detail, and actual contradictions occur occasionally. In such a case, it is not
clear how such contradictions are to be resolved, on the basis of the publication
alone.

TAXONOMIC DATA MATRIX

The starting point for many applications of computers to taxonomy is the data
matrix. This is a rectangular table by taxa in one direction and characters in the
other, with entries (character states) filled in numerically. Such a matrix can

represent any level of taxon, such as species, genus and family, and any kind of character, such as qualitative or quantitative, with two or many states.

A taxonomic data matrix is the basis for methods of identification by computer, such as key or polyclave construction, on-line identification systems or matching programs. It can also be used, after similarity (or dissimilarity) coefficients have been calculated from it, for any of very many numerical clustering techniques.

Numerical reconstructions of phylogenies (e.g. Estabrook, 1968) require a taxonomic data matrix as part of the data. The matrix can also be used for searching for correlations between characters, or with other information, such as observations on climate.

INCOMPLETE DATA AND COMPUTER IDENTIFICATION

The effect of incomplete data on identification by computer is discussed. The effects of inconsistent descriptions are similar because if the inconsistencies cannot be resolved, then gaps in the data matrix result.

The construction of keys is hampered by missing data because it is difficult to construct opposing branches in a key with a particular character unless that character is known for each of the relevant taxa. Unknown characters could be filled in on the assumption that the taxon is capable of showing any of the relevant character states, but this results in some taxa appearing more than once in the key, and the key is lengthened and is less effective. The same situation occurs with characters which vary naturally, and hence constant characters are preferred for key construction, but with missing characters there is always a risk that in fact the taxon exhibits some state which is not observed in the others. Experience with constructing keys by computer (about 50 to date) from data derived from conventional publications suggests that the published data are often only just sufficient for constructing a key, and leave little room for constructing keys in different ways. Alternatively, if the data are insufficient for constructing any key, then only a partial key can be produced, i.e. a key in which not all taxa can be separated.

A polyclave can always be produced, with missing data, by pretending that missing characters exhibit all possible states. This then means that characters with missing data fail to differentiate as many taxa as they might otherwise do, and the polyclave is then less effective. The same objection about unknown character states still applies, as in keys.

With the matching method for identification (Pankhurst, this volume, Chapter 6) the effect of missing data is once again to degrade the performance of the program. The data on *Rubus* which were used to test this program were

derived initially only from the monograph, and the program frequently failed
to include the correct determinations in its output. Experiments then showed
that the best way to improve its performance was to complete the matrix as far
as possible, and this was done, with marked benefit. The reason for this was that
random errors were introduced into the estimation of similarity by the missing
data, since similarities between particular pairs of taxa were based on different
sets, and different total numbers of characters than were other pairs.

The effect of missing characters on automatic identification techniques then
appears to be generally a degrading one. There is evidence that similar effects
occur with similarity measures used for clustering (Crovello, 1968).

EXAMPLE OF DATA QUALITY

The data contained in a monograph of the British members of the genus *Rubus*
(Watson, W. C. R., 1958) are examined in detail. From this work a data matrix
for 161 characters on 389 taxa was constructed. This was in fact done with the
aid of a computer system for data capture (Pankhurst, 1972). There is little
evidence available to show whether this particular monograph is better or worse
than any other for the quality of its data.

Table I gives an account of the distribution of the information about various

TABLE I. Distribution of character information in Watson's Rubi

Proportion of taxa for which character is recorded	No. of characters
90–100%	11
80–90%	8
70–80%	4
60–70%	11
50–60%	10
40–50%	11
30–40%	20
20–30%	21
10–20%	19
up to 10%	46 (305)[a]
	161 (420)

[a] *Note.* About 420 characters are mentioned altogether, and if this number of characters
is used, then the total number of characters in the last class (up to 10%) is increased to 305.

characters in terms of how many taxa have this information recorded. No one character is recorded for each taxon throughout the whole work. This is a surprising fact, and it must be presumed that neither the author nor the editors checked this point (the author died before the work was edited). For 389 taxa, the theoretical minimum number of characters required to separate all taxa from each other in the ideal case is $\log_2 389$, or approximately 9. This figure has to be multiplied by some factor to allow for correlation between characters. An experiment with subsequently completed data gave 27 characters as an estimate of the actual minimum number. This should be compared with the figures in Table I. In other words, the data in this monograph are probably inadequate for the construction of any complete key.

The average number of characters scored for a taxon was 56, with a minimum of 24 and a maximum of 86. From this one can quickly calculate the percentage completion of the matrix

i.e. $$\frac{56}{161} \times 100 = 35\%.$$

Some of the characters are conditional on other characters, so this must account for part of the missing 65%. A careful survey showed that 15 characters were conditional. If it is assumed that these characters are unobservable half the time, then $\frac{15}{161} \times \frac{100}{2} = 5\%$ are impossible to observe, which does not at all account for the missing data.

Inconsistencies in the description of states of characters totalled about 13 out of 161. It is difficult to estimate what percentage of the matrix is affected by this, but this figure is not high. The monograph included descriptions of taxa of intermediate rank, as well as a partial key built into the text. A computer program was used to detect discrepancies between this supplementary data and the text proper, and about 40 disagreements were observed. This means about one error of this kind per ten descriptions, or a rate of about 1 in 550 of characters recorded. This figure is probably about average in human clerical work, but it is distressingly high when all the other kinds of undetectable errors which must occur, such as mis-observation, are considered.

COMPLETING THE MATRIX

Experience with the problems connected with completing a data matrix are described here. Watson's Rubi are again taken as the example. The only way to augment this particular matrix was to obtain preserved material and re-describe it, since there exist no living collections. The original author could not be consulted since he is deceased. His original collections were not of good quality, and are scattered in various institutions, and some have been destroyed. This situation is probably quite typical, and the only way to proceed was to use the collections of B. A. Miles (an associate of Watson) which are of high quality, well annotated, and based on more or less the same taxonomic concepts.

A selection of voucher specimens was taken from a herbarium.* Luckily all but about 10 out of 400 species were represented. Two or three specimens were chosen in each case, as were judged typical by comparison with the others. It is not suggested that this number of vouchers is adequate, but that it was only practical to deal with a few specimens of each to begin with. All voucher specimens were numbered and marked, and noted in a card index.

The characters to be scored were chosen on the basis of those which were most frequently recorded in the monograph, and which were also visible on dead specimens or noted on the labels. A great deal of care was taken to define exactly what was meant by each character, and this was set out in a report (Pankhurst and Aitchison, 1973).

The characters were recorded on punched paper tape using a programmable paper tape punch machine (Friden Flexowriter). The characters were recorded species by species in small groups of about five characters at a time. The species were examined in the same order every time, and all available states of the current set of characters punched. It was found to be convenient to punch a separate paper tape to control the punching machine, as has been reported before (e.g. Soper and Perring, 1967). A program was also written to punch the control tapes. The input to the program was a list of desired characters and a list of all the species names. The output tape, when put on the punch machine, typed the name of each species in turn, and punched only a code number for it, and then proceeded to type the name of the next character wanted, while punching a code for the character, and then stopped. The operator then typed in integer code numbers for the state or states observed (if any) for this character on this species, and pressed a "start" button. The machine then went on with the next character, and eventually to the next species. In this way, a paper tape was obtained, containing the observed characters in a condensed and coded form, ready for further machine processing. It may be noted that data capture by this means is only convenient when the characters can be presented in a fixed order as with preserved material. An attempt to do this with descriptive text would be unsatisfactory because the characters are usually not presented in an absolutely constant sequence, which would make necessary laborious scanning of text.

The combination of the data from the monograph and that newly observed

---

* University of Cambridge, **CGE**.

presented several problems. Firstly, it was necessary to check with Watson's descriptions that the voucher specimens were indeed correctly identified. This was done in several stages. The matching program (Chapter 6), based on the monograph data alone, was presented with the voucher specimens as unknowns. All vouchers which agreed by 70% or more were accepted. Others were examined in greater detail, using the complete original text and taking note of the italicized characters. Adequate agreement here was taken to mean 70% or more with a hand-made character count. After this, some of the vouchers remained, which did not agree sufficiently, and these were rejected. In this case, data from the vouchers was not included in the combined matrix.

Secondly, it was necessary to make sure that the new data were based on the same definition of characters as used by the original author. In a few cases, the author defined his terms, but mostly he did not. An appeal to common usage, such as set out by Stearn (1973) does not cover all cases, and so experiments were made to test whether the character definitions were correct. This was done on the basis of comparing the old and new data character by character. All characters which agreed by 70% or more over the same set of species were taken to be correct. A program was written to make these comparisons. Some cases remained where agreement was inadequate, and here only the new observations were retained, on the grounds that it is now impossible to discover what the original author meant. An example of this was the character "petal width". The author used the terms "narrow" or "broad" (omitting "moderate"), and did not state whether he used absolute width or width : length ratio, nor did he state the numerical limits. It is possible that he simply estimated this character visually without taking measurements. The character is now defined to mean the ratio, with specified limits.

The combined data from the two sources gave a new data matrix which is 91% complete. The missing 9% includes the effect of conditional characters, and also missing data caused by those species without voucher specimens or with inadequate ones.

CRITERIA FOR TAXONOMIC DESCRIPTIONS

From the preceding discussion it is clear that, in order to make the most of certain computer techniques in taxonomy, new standards of taxonomic description will be needed. It will now be assumed that this is desirable, and the implications will be discussed.

Two main criteria are necessary: (1) the data must be as complete as possible; and (2) the character descriptions must be consistent, i.e. the states applying to one character must be mutually exclusive. It is probable that these criteria have long been regarded as desirable by some, but that without the use of computer methods in collection of data, they could not easily be achieved.

From this standpoint, one can regard conventional taxonomic descriptions as faulty, and these faults will now be discussed in general terms. Although descriptions are often written to serve for identification it is rare that any explicit theory of how this is to be done is set out, except with regard to diagnostic keys.

In the case of keys, it is desirable that the key and the corresponding text should agree exactly. That minor errors should occur from clerical mistakes is understandable, but surprisingly there is evidence that authors are not always of the opinion that keys and text ought to agree. Some authors for Flora Europaea checked keys and text for agreement, and some required that no extra data appear in the keys, while others did neither (Walters, S. M., personal communication, 1972). Another point of confusion about keys is whether they should follow the principles of the classification (a conspectus or "natural key") or whether they should simply use whatever characters are most effective, and if necessary, group taxa without regard to their classification (an artificial key). In general, a key used to demonstrate the classification is less effective for identification, because of the exceptions and qualifications about the characters. Hence an artificial key is best for identification and a conspectus or "natural key" is a way of setting out the classification, and these two purposes ought to be kept distinct.

After a key has been used and a tentative identification obtained, it is customary to refer to the text (among other things) for confirmation. The description should then agree with the specimen, more or less. No guidance is generally offered as to how much agreement is necessary before a positive identification is made, but at a guess, one might expect eight out of ten characters to be correct. In addition, particular character states are often italicized, and this may be taken to mean that these characters are more important, and presumably if more than a few of these "special" states disagree, one will tend to reject the identification. This mental procedure can be compared with a computer program for matching.

A common fault in this approach is that the author gives bias to the information which is used for identification by supplying little more than those facts that are necessary to carry out the identification in one particular way only. Consequently any reader who may have sound practical reasons for wanting to approach the identification in a different way, finds it very hard to do so. There is also a tendency for this situation to perpetuate itself through successive taxonomic revisions. An example of this occurs in a treatment of the genus *Ononis* (Ivimey-Cook, 1968), where the key begins, by tradition, with an unsatisfactory character, and where the author has attempted, without success, to escape from this.

A solution to some of the above difficulties would be to approach the writing of a handbook by first constructing the data matrix, and then subsequently to construct the keys and the descriptive text. Several programs are currently available which will do both these things simultaneously (with judicious editing) and ensure consistency. This is in fact rather like the approach adopted by the Flora North America (Shetler *et al.*, 1973) and the Flora of Veracruz (Gómez-Pompa and Nevling, 1973a).

Objections can be made to the above criticisms. One of these is that the effort required to complete a data matrix is prohibitive. One must distinguish here between the problem of getting at the sources of information (living organisms, collections and publications) as opposed to actually extracting the data. If one may assume, as is usual in monographic work, that it is essential to make collections from the wild and study living material, then the author will normally examine a large part of the relevant material as a matter of course. In that case, one may ask how much additional effort would be required to complete the descriptions. This can be estimated in the case of Watson's Rubi. Assume that characters can be recorded, by examination of several voucher specimens at once, at the rate of two a minute. Compared with my own experience, this is a conservative figure. Assuming a 40 h week, this comes to a total of about 12 weeks' work. Watson did about a third of this, so the remainder is 8 weeks. This is apparently a great deal of effort, except when one realises that Watson devoted 40 years to this study as a whole. This must also be compared with the effort required to retrieve these data again at a later stage, with the extra attendant difficulties, which in this instance was approximately 6 man-months, even with use of computer aids (Pankhurst, 1972). If a handbook is being compiled from other publications, without substantial new research, as in Flora Europaea, then the objection is more critical, but even here much extra data could be obtained by reference to one good collection, even if the law of diminishing returns makes it not worthwhile to search out the remainder elsewhere.

Another objection is that complete descriptions would be too long to put in print. This can be met by several arguments. One is that suitable heading notes before descriptive text could save much space. For instance, one could have statements such as: "All species have white petals unless otherwise stated". In the Watson example, much effort could have been saved in the re-discovery of data if this device had been adequately used. Otherwise, one is obliged to assume that the author had not observed many of the rather common states of characters, and then re-observe them in order to be sure. Another argument, perhaps more important, is that even if it makes no sense to publish all the information, it is of great benefit to store it, for those who will need it later. This remains true whether office or computer methods of storing data are used. It is likely that there has been immense wastage of effort over the years because taxonomists have not documented their research in such a way as to make it easy for others to take up afterwards. This applies not just to descriptions but also to all kinds of other information, such as bibliographic sources, location of wild specimens and of type material, and notes on which specimens have been used for preparing descriptions.

CONCLUSION

Methods of automatic identification will become much more effective if two standards, those of (a) completeness and (b) consistency, are applied to the establishment of taxonomic descriptive data. Completeness can be achieved by proceeding much as in the past, but by being rigorously methodical. The extra labour involved in this may not be large, relatively speaking, and can be reduced by using computer-assisted data capture. Consistency can be achieved for groups of closely related organisms by a little forethought, but for comparison of more diverse taxa, a standard set of descriptive terms is needed. This has already been attempted by the Flora North America programme (Shetler *et al.*, Chapter 13, this volume, and 1973) and recording of standardized descriptive data is currently being carried out, on a smaller scale, in the Flora of Veracruz (Gómez-Pompa and Nevling, 1973b). It is possible that standards for the contents of data matrices for use in computing processes will be more important than standards for the format of such data.

REFERENCES

CLAPHAM, A. R., TUTIN, T. G. and WARBURG, E. F. (1962). "Flora of the British Isles" (2nd edition). Cambridge University Press.
CROVELLO, T. J. (1968). The effects of missing data and of two sources of character values on a phenetic study of the willows of California. *Madroño* **19**(8), October 1968.
ESTABROOK, G. F. (1968). A general solution in partial orders for the Camin-Sokal model in phylogeny. *J. theor. Biol.* **21**, 421–438.
GÓMEZ-POMPA, A. and NEVLING, L. I. Jr. (1973a). The use of electronic data processing methods in the Flora of Veracruz program. *Contr. Gray Herb.* **263**, 49–64.
GÓMEZ-POMPA, A. and NEVLING, L. I. Jr. (1973b). Ordenación de Datos para la Descripción de Especies para la Flora de Veracruz. *In* Flora of Veracruz, Publication no. 1, pp. 34–41. University of Mexico.
HEYWOOD, V. H. (1973). Ecological data in practical taxonomy. *In* "Taxonomy and Ecology" (V. H. Heywood, ed.), pp. 329–347. Academic Press, London and New York.

IVIMEY-COOK, R. B. (1968). Genus *Ononis*. *In* Flora Europaea, Vol. 2, pp. 143–148. Cambridge University Press.

PANKHURST, R. J. (1972). A method for data capture. *Taxon* **21**(5/6), 549–558. November 1972.

PANKHURST, R. J. and AITCHISON, R. R. (1973). Notes on recording the descriptions of *Rubus*, Stage 3. Internal report no. 4, Identification Project, Botany School, Cambridge.

SHETLER, S. G. *et al.* (1973). A guide to contributors to Flora North America, Smithsonian Institution, Washington. Part I (FNA Report 65) and Part II (FNA Report 66).

SOPER, J. H. and PERRING, F. H. (1967). Data processing in the herbarium and museum. *Taxon* **16**, 13–19, February 1967.

STEARN, W. T. (1973). "Botanical Latin" (2nd edition). David and Charles, Newton Abbott.

WATSON, L. (1971). Basic taxonomic data: the need for organization over presentation and accumulation. *Taxon* **20**(1), 131–136.

WATSON, W. C. R. (1958). "Handbook of the Rubi of Great Britain and Ireland". Cambridge University Press.

Statistical Theory

15 | Relating Classification to Identification

J. C. GOWER

Rothamsted Experimental Station, Harpenden, Hertfordshire, England

Abstract: Multivariate mixture and discriminant problems are classical statistical and identification problems which are shown to be closely interrelated. This is just one example of when the rigid distinction between classification and identification becomes blurred. Another example is given relating non-probabilistic classification and identification. It is suggested that these are not isolated cases, but that classes are often set up so that future samples can be assigned to them optimally.

Key Words and Phrases: classification, clustering, grouping, population-mixtures, identification, diagnosis, assignment, discrimination, keys

INTRODUCTION

This contribution discusses identification problems with special reference to statistical work. It seemed to me that a general contribution along these lines would provide an interesting background against which the more specialist topics, with which this volume is mostly concerned, could be usefully assessed. Payne (this volume, Chapter 4) discusses our joint work on diagnostic keys.

In recent years problems of assignment to classes (i.e. identification) have been carefully distinguished from problems of constructing classes. It will be one of my aims to show how intimately the two can be related.

HISTORICAL BACKGROUND

The first statistical paper on what would now be regarded as a classification problem, seems to have been written by Karl Pearson (1894). He was concerned with resolving a set of observations into two components or classes, that is, the data were regarded as being a mixture of values drawn from two separate biological populations. Only a single variate was concerned and each component of the mixture was assumed to be normally distributed. The object was to

Systematics Association Special Volume No. 7, "Biological Identification with Computers", edited by R. J. Pankhurst, 1975, pp. 251–263. Academic Press, London and New York.

estimate the parameters of both populations and the proportions of each in the total sample. A typical example of this kind of problem occurs when one has measurements of fossil (or ancient) bones which are presumably a mixture from male and female sources. One may wish to estimate average lengths etc. for males and females separately, and a subsidiary requirement may be to assign (or identify) each bone to one or other of the sexes. Figure 1 illustrates a mixture sample with fitted normal distributions. The vertical line separates samples which will be assigned to the different components and shows that some mis-allocations are inevitable. As the means of the two components get closer, mis-allocation worsens.

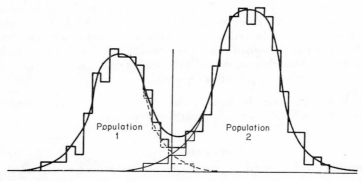

FIG. 1. A histogram of a sample from a mixture of two normal populations. The heavy line shows the combined sample. The light continuous and dotted lines indicate contributions from the individual populations that cannot be distinguished. Samples falling to the left of the vertical line will be assigned to population 1, and those to the right, to population 2.

Mixture problems of this kind have been intermittently studied over the years but only recently have multivariate forms of the problem begun to receive attention. This is because even the univariate case poses formidable computational and theoretical problems. The multivariate case can scarcely be tackled without modern computers and the properties of the estimates found are little understood, but it would be surprising if they were better than those in the univariate case; indeed it is likely that they are worse. One formulation of cluster analysis as a class-construction method is a direct descendant of Karl Pearson's mixture problem, emphasizing the assignment part of the analysis. Working in terms of multinormal mixtures, this is the approach taken by Wolfe (1970), and is closely associated with the work of Friedmann and Rubin (1967) and John (1970a). Day (1969) and Scott and Symons (1971) have also made useful contributions here. John (1970b) has begun to explore mixtures of other kinds of multivariate distributions. The formulation of the equations for estimating the mixture parameters is straightforward but computation is difficult. If cluster analysis is treated as a mixture problem, other more simple computational processes may be viewed as non-parametric estimation procedures and studied for their efficiency.

Fisher (1936) initiated another line of statistical development which he termed discriminant analysis. In discriminant analysis, samples are available from known biological populations and the problem is how to combine different measurements on future samples so that they can be identified most efficiently. In his 1936 paper, Fisher's example was the famous one, of how to assign an iris to one of three populations, and was based on four measurements—sepal length and width, and petal length and width. Fisher was specifically concerned with finding the best *linear* function of these measurements for identification (or discrimination) purposes. "Best" can be interpreted in many different ways and Fisher used as a criterion a concept drawn from his development of the analysis of variance, namely, the linear function that maximizes the ratio of the between-population variance to the within-population variance. Welch (1939) using a different criterion based on minimizing probabilities of mis-classification of future samples showed that a linear function was the best discriminator for multinormal populations with the same covariance matrices. With Welch's criterion other types of population would require non-linear discriminators. Fisher's work had been foreshadowed in a paper by Barnard (1935) which was concerned with discriminating between four groups of Egyptian skulls, but Fisher himself had been largely responsible for the methods used in that paper.

Thus, by the 1950s probabilistic classification problems were well established in statistical circles and were generally understood as being concerned either with separating mixtures, together with assigning samples to particular mixture components, or with developments of Fisher's discriminant functions which were also concerned with identifying samples. Thus the term classification was more concerned with identification than with constructing classes.

The increasing availability of computers in the late 1950s stimulated an interest in their possible uses in taxonomy and related sciences. Working independently, Sneath (1957), Sokal and Michener (1958) and Williams and Lambert (1959), with interests respectively in bacteriology, entomology and ecology produced computer programs for classificatory purposes and mostly for hierarchical classification. The immediate effect of this work was two-fold. Firstly, it stimulated widespread interest and encouraged many others to develop similar

computer programs. Secondly, it initiated a period of general confusion which
is only now being resolved. It was not immediately realized by all that the
purpose of this work was to construct classes rather than to help taxonomists
assign individuals to previously established classes, and this confusion was
especially strong in statistics, because statisticians had long associated the word
classification with assignment. Once statisticians became aware of this distinction
they were careful to emphasize in future writings whether they were concerned
with an assignment or a classification problem; thus the term classification then
began to be reserved for ideas of constructing classes and the terms assignment,
diagnosis and identification used for the older type of discriminant problems.
However, the words classification and discrimination are still used interchangeably, especially in the more mathematical literature.

Biologists were much concerned with the connection between phylogenetic
classifications and the hierarchical classifications found by the new methods. It
was quickly asserted that computer-found classifications were not intended to be
interpreted in an evolutionary context, but merely reflected the similarities
between populations as assessed from their observed morphological or other
characteristics. The controversy still continues, and the evolutionists have a valid
point when they ask that if the hierarchies are not estimates of phylogeny and
usually cannot be used as diagnostic keys, because they are polythetically based,
what use are they? Recently more direct attempts have been made to construct
evolutionary trees but these need not detain us here (see Edwards, 1970;
Farris, 1972).

A source of concern to statisticians (remarkably close to the criticisms of
evolutionists) was that the new methods were presented in purely algorithmic
forms which obscured the properties of the classes found. Classes can be constructed for many purposes and objective criteria for some purposes have been
explored in recent years. This work provides a much firmer basis for
classificatory cluster analysis and its ramifications, and I shall return to it
shortly.

Since the 1950s, work on constructing classes has grown rapidly, but, apart
from highly technical studies of discrimination error rates (see below) in the
statistical literature, corresponding identification problems have had relatively
little emphasis. The special problems of medical diagnosis and pattern
recognition, which have close links with statistical discrimination, have received
considerable attention, but not ordinary biological taxonomic identification.
There has been a sprinkling of papers which has grown within the last 2 or 3
years. Perhaps we are on the verge of a deluge similar to that experienced with
classification problems.

DISCRIMINATION

Suppose we have two biological populations with multivariate frequency functions $f_1(\mathbf{x}, \boldsymbol{\theta}_1)$ and $f_2(\mathbf{x}, \boldsymbol{\theta}_2)$. To begin with we shall assume that the parameters $\boldsymbol{\theta}_1$ and $\boldsymbol{\theta}_2$ are known. Suppose now that we have a rule R that assigns a sample to one of the two populations when measurements of $\mathbf{x}$ are available. Then, four outcomes are possible:

    (1) an individual from population 1 is correctly assigned to population 1
    (2) an individual from population 2 is correctly assigned to population 2
    (3) an individual from population 1 is incorrectly assigned to population 2
    (4) an individual from population 2 is incorrectly assigned to population 1
the probabilities of which may be written

$$p(1, 1) \; p(2, 2) \; p(1, 2) \; p(2, 1).$$

The quantities $p(1, 2)$, $p(2, 1)$ are known as probabilities of mis-classification and the object of discriminant analysis is to determine the rule R so that some function of the probabilities of mis-classification is minimized. Figure 2 identifies geometrically these probabilities for two univariate normal populations. It will be seen in the next section that the similarity between Figs 1 and 2 is not fortuitous.

Two criteria are often studied:

    (1) minimize $p(1, 2) + p(2, 1)$, the total probability of mis-classification.
    (2) minimize Max $(p(1, 2), \; p(2, 1))$, the maximum probability of mis-classification.

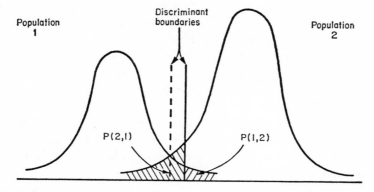

Fig. 2. Discrimination for two normal populations and one variable. If the discriminant boundary is taken as the solid vertical line, the errors of misclassification will be proportional to the shaded area. This is least when the boundary is drawn through the point of intersections of the two curves (dotted line).

Clearly this formulation extends without difficulty to several populations, the aim being to divide sample space into regions in which samples are assigned to populations in such a way that probabilities of making incorrect assignments are minimized according to some criterion. The second criterion above is known as the minimax rule. It implies that all probabilities of mis-classification are equal, for if not, we can always adjust the discriminant boundaries so as to decrease the maximum probability of mis-classification at the expense of increasing the smaller ones (see Rao, 1952, p. 313). As Rao pointed out, it is unreasonable to give equal probabilities of mis-classification for three populations when two are easily confused populations, while a third population is very different. For this kind of reason, and also because of computational difficulties, the minimax criterion is not much used for discrimination.

A consideration that is often relevant when calculating the effects of mis-classification is whether or not to allow for costs of different kinds of mis-classification and for population frequencies (usually referred to as their *a priori* probabilities). If these costs and frequencies are known, it is simple to allow for them, and in commercial and social contexts this may be important. For example in English law it is usually considered better to free a guilty man than to convict an innocent man, implying, in a general sense, that different costs are assigned to the two outcomes. But in many taxonomic situations costs are irrelevant and it may be unreasonable to allow the relative rarity of certain species to give adverse weight to the probabilities of mis-classification when these rare species are in fact found.

Several mathematical derivations of the discrimination rules can be found in standard textbooks and I shall not give a detailed discussion here. Anderson (1958) gives a good account of what are termed Bayes procedures and Rao (1952) concentrates more on the likelihood-ratio approach. Fisher's original variance-ratio method is also discussed in these (and more recent) textbooks but it is now recognized as being a special case that is optimum only for multi-normal populations. All these approaches lead to essentially the same results, that are summarized below.

Denoting the *a priori* probability of the ith population by π_i and the cost of incorrectly assigning a sample from the ith population to the jth, by $C(i, j)$ then the rule that minimizes the expected cost of mis-allocation is to assign a sample $\mathbf{x}$ to the kth populations if:

$$\sum_{i \neq k} \pi_i f_i(\mathbf{x}, \boldsymbol{\theta}_i) C(i, k) < \sum_{i \neq j} \pi_i f_i(\mathbf{x}, \boldsymbol{\theta}_i) C(i, j)$$

$$\text{for } j = 1, \ldots, m; \quad j \neq k.$$

The left-hand side is merely the expected cost of mis-classifying an individual from the kth population and the right-hand side is the expected cost of mis-classifying an individual from the jth population. The rule merely says that we have to work out the expected cost of mis-classifying the sample if it belongs to the jth population taking $j = 1, 2, ..., m$ in turn. The value of j ($=k$, say) for which this cost is least gives the best choice of population. Stated this way, the result is almost obvious and many discrimination results, stripped of their algebraic dress, are of this kind. When the *a priori* probabilities and costs can be ignored the rule is to assign to population k if

$$\sum_{i \neq k} f_i(\mathbf{x}, \boldsymbol{\theta}_i) < \sum_{i \neq j} f_i(\mathbf{x}, \boldsymbol{\theta}_i)$$

for $j = 1, ..., m; \quad j \neq k,$

which states that the sum of the ordinates of the frequency functions for all populations, except the kth, is to be less than for any other choice of k. Since $\sum_{i=1}^{m} f_i(\mathbf{x}, \boldsymbol{\theta}_i)$ is constant, the above is only a complicated way of writing

$$f_k(\mathbf{x}, \boldsymbol{\theta}_k) > f_j(\mathbf{x}, \boldsymbol{\theta}_j) \quad j = 1, ..., m, \quad j \neq k.$$

That is one assigns $\mathbf{x}$ to the population whose frequency function, or density, is maximum for the observed value $\mathbf{x}$.

The advantage of this theoretical approach is that it is derived quite independently from any assumptions about the exact distributional forms. Of course, in applications these forms must be specified and when substituted into the formulae may give computational problems, but this is a numerical rather than statistical problem. Although in many taxonomic situations quantitative and qualitative variables are observed and there is insufficient knowledge to postulate the multivariate forms $f_i(\mathbf{x}, \boldsymbol{\theta}_i)$, it is far from true, as is sometimes thought, that discriminant analysis is restricted to multi-normal populations.

Figure 3 shows curvilinear boundaries discriminating between three normal populations with differing dispersions.

In practice, estimated values $\hat{\boldsymbol{\theta}}_i$ of the parameter $\boldsymbol{\theta}_i$ have to be used and their use detracts from the optimum properties of the discriminators, because the probabilities of mis-classification $p(i, j)$ now depend on the estimates $\hat{\boldsymbol{\theta}}_i$ and not the unknown $\boldsymbol{\theta}_i$. Much mathematical work has been concerned with the calculation of the distributions of the estimated $p(i, j)$ especially when the $f_i(\mathbf{x}, \boldsymbol{\theta}_i)$ are assumed multinormal.

Another problem is how to cope with samples that may not be from any of the given populations. A solution is not to divide the whole sample space into regions associated with definite populations, but to allow for *doubtful regions* with no assignment.

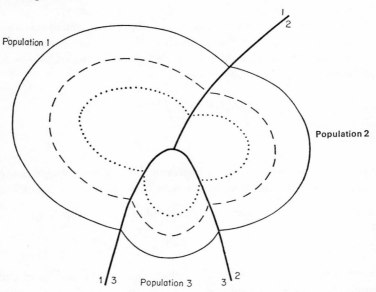

Fig. 3. Hyperbolic discriminant boundaries (bold lines) between three bivariate normal populations with unequal dispersion matrices.

MIXTURE PROBLEMS

Suppose we have k populations with frequency functions $f_1(\mathbf{x}, \boldsymbol{\theta}_1), f_2(\mathbf{x}, \boldsymbol{\theta}_2), \ldots,$ $f_k(\mathbf{x}, \boldsymbol{\theta}_k)$ and a sample $\mathbf{x}_1, \mathbf{x}_2, \ldots, \mathbf{x}_n$ of size n from which we wish to estimate $\boldsymbol{\theta}_1, \boldsymbol{\theta}_2, \ldots, \boldsymbol{\theta}_k$ and assign each of the n samples to one of the k populations. The likelihood for this sample is

$$L = \Pi_{\mathbf{x}\varepsilon C_1} f_1(\mathbf{x}, \boldsymbol{\theta}_1) \, \Pi_{\mathbf{x}\varepsilon C_2} f_2(\mathbf{x}, \boldsymbol{\theta}_2) \, \ldots \, \Pi_{\mathbf{x}\varepsilon C_k} f_k(\mathbf{x}, \boldsymbol{\theta}_k)$$

where C_i is the set of observations assigned to the ith population. Suppose the assignments have been made so that the likelihood L is maximized, then moving a sample $\mathbf{x}$ from C_i to C_j reduces the likelihood, giving

$$L \leqslant L f_j(\mathbf{x}, \boldsymbol{\theta}_j)/f_i(\mathbf{x}, \boldsymbol{\theta}_i)$$

i.e. $$f_i(\mathbf{x}, \boldsymbol{\theta}_i) \leqslant f_j(\mathbf{x}, \boldsymbol{\theta}_j).$$

Thus with a maximum likelihood formulation of the mixture problem the boundary between samples assigned to populations C_i and C_j is exactly that given by the likelihood ratio discriminant regions discussed previously. In fact the above can be interpreted as giving an alternative derivation of the classical discriminant rules. When the quantities θ_i also are to be estimated, the correspondence is no longer exact, but neither then is the sampling theory for discriminant functions. This shows that the classes $(C_1, C_2, ..., C_k)$ formed, may be regarded as having been constructed to optimize assignment.

If the population frequencies $\pi_1, \pi_2, ..., \pi_k$ are also to be estimated, then the new likelihood is $L' = L\Pi_{\pi_i} \mid C_i \mid$ where $\mid C_i \mid$ is the number of samples assigned to the ith class, so that the discriminant regions found above are modified by the population frequencies exactly as for the discriminant theory outlined in the previous section.

This is a fairly mathematical example of where classification and identification are closely related. In the next section I show how similar properties apply in non-probabilistic classification problems close to those of classical taxonomy.

MAXIMAL PREDICTIVE CLASSIFICATION

This form of classification is a mathematical formulation of ideas put forward by Gilmour (1937, 1940, 1951, 1961), that embody his well known dictum that "a system of classification is the more natural the more propositions there are that can be made regarding its constituent classes". In its simplest form we have n populations to be classified into k classes on the basis of v dichotomous characters. Probabilistic considerations are no longer relevant because we shall consider that the characters have been selected so that they do not vary within the populations. This is a somewhat idealized situation, but is not an unreasonable model for major taxonomic classifications and is an implicit assumption when constructing hierarchic keys. It is only closely related populations that overlap, making probabilistic considerations inescapable.

With a single class a list of the states of the v characteristics that are most common within that class can be made and used to predict the properties of the class. The number of correct predictions within the class is then the number of times the most common characteristics occur. The list of properties predicted for a class is termed a *class predictor*, and may be identified with Gilmour's "propositions".

Table I shows a class of four populations with five characters together with its class predictor.

The predictor for the fifth character could be either 0 or 1 because both states are equally frequent. In either case there are 14 correct predictions.

Table I. A class of four populations with five characters

Character	I	II	III	IV	V
Population A	1	1	0	1	1
B	1	1	1	0	0
C	1	0	0	0	0
D	0	1	0	0	1
Class predictor	1	1	0	0	0/1

With k classes there are k class predictors and the number of correct predictions within each class can be summed. Table II shows 11 populations divided into three classes.

There are respectively 11, 14 and 14 correct predictions in the three classes giving a total of 39 for this particular classification.

TABLE II. Eleven populations divided into three arbitrary classes

Character	I	II	III	IV	V
Population A	0	1	0	0	1
B	1	0	0	1	1
C	1	1	0	0	0
Class 1 predictor	1	1	0	0	1
Population D	0	1	1	1	0
E	1	1	0	0	0
F	0	0	0	0	0
G	1	1	0	1	0
Class 2 predictor	0/1	1	0	0/1	0
Population H	0	0	0	0	0
I	1	0	0	0	1
J	0	0	0	1	1
K	1	0	1	1	1
Class 3 predictor	0/1	0	0	0/1	1

The maximal predictive classification occurs when this sum is the maximum attainable for all possible ways of classifying the populations into k classes. Gower (1974) has discussed this type of classification in detail. Here I am only concerned with the identification properties of classes produced in this way.

Clearly any individual cannot have fewer properties in common with its own class predictor than with the predictor of another class C (say), because if it did the classification would not be maximal, as the number of correct predictions can be increased by transferring the individual to C. Thus to identify an individual, its v characteristics need be matched only against the k class-predictors and not against all n populations. The class-predictor giving most matches must be that of the class to which the individual belongs. In this matching one should count all comparisons with an equi-frequent 0/1 predictor as a match. Usually a unique class-predictor will give most matches but in pathological situations more than one may be found. In this way an individual is assigned to a subset of populations which must contain the one being identified. The process can be continued by forming maximal predictive classes from the subsets. When k is chosen to be 2 at every step, we have something like a binary diagnostic key, except that each node is associated with a set of polythetic tests rather than a single monothetic test.

The above process is clearly related to the use of matching-coefficients, and I have suggested elsewhere (Gower, 1974) that maximal predictive classes lie behind much of the modern development of classification and hierarchical classification based on taxonomic distances and similarity. It is interesting to see how a clearly stated criterion for classification has again led to classes with optimal identification properties.

As with discrimination and mixture problems the *a priori* probabilities of the different populations can be introduced, without any material complication. Furthermore differing probabilities of character-states within a population cause no difficulties so long as the same combination of states cannot occur in two or more populations, i.e. there is no overlap. When populations do overlap a full probabilistic treatment is necessary. Even then classification and identification can be linked, as the following example shows. Suppose we have n (probabilistic) populations which are to be arranged into k classes. One possible criterion is to choose the boundaries between the classes so as to minimize probabilities of mis-classification, just as in discriminant analysis.

CONCLUDING REMARKS

Much of the above has no direct bearing on the empirical algorithms that have so far been used for constructing diagnostic keys for biological populations with non-probabilistic characters. However, I think it useful to view this work within the broader context given in this paper. We are generally concerned with assigning to populations or groups of populations and are not greatly concerned with single item populations that occur in some non-biological fields. A rapid survey of statistical assignment has focused attention on the distinction between probabilistic and non-probabilistic populations and the different criteria appropriate for assignment to them. Biologically close populations can only be described with variable (i.e. probabilistic) characters but distant populations often do not have this drawback. Failure to recognize the special properties of probabilistic and non-probabilistic data often leads to inappropriate analyses especially as both can be presented in $n \times v$ matrix form that cannot be distinguished by a computer program. An extreme type of failure occurs when each set of probabilistic populations is represented by a single sample, giving a spurious impression of invariability within populations.

Although I have been concerned to show the relationship between some forms of classification and identification, it has not been my intention to imply that the two are necessarily related. Many other objective criteria for classification could be stated which would not necessarily help with identification. Nevertheless an important reason for classifying things is so that one can name and then talk about the classes, conveying as much information as possible in the process and this often involves being able to assign things efficiently to the named classes. Indeed it can be argued that classification is a cornerstone of all language, and of scientific language in particular. Boldrini (1972) writes:

"The verbal reality of the common noun, as well as of the propositions in which something is predicted of it, and of the arguments and scientific structures of which, in a certain sense, it becomes the protagonist, give to modern science the appearance of a discourse: and it is therefore said that scientific reality is language."

REFERENCES

ANDERSON, T. W. (1958). "An Introduction to Multivariate Statistical Analysis". Wiley, New York.

BARNARD, M. M. (1935). The secular variation of skull characters in four series of Egyptian skulls. *Ann. Eugen.* **6**, 352–371.

BOLDRINI, M. (1972). "Scientific Truth and Statistical Method". Griffin, London.

DAY, N. E. (1969). Estimating the components of a nature of normal distributions. *Biometrika* **56**, 463–474.

EDWARDS, A. W. F. (1970). Estimation of the branch points of a branching diffusion process. *J. R. statist. Soc.*, B **32**, 155–174.

FARRIS, J. S. (1972). Estimating phylogenetic trees from distance matrices. *Am. Nat.* **106**, 645–668.

FISHER, R. A. (1936). The use of multiple measurements in taxonomic problems. *Ann. Eugen.* **7,** 179–188.

FRIEDMAN, H. P. and RUBIN, J. (1967). On some invariant criteria for grouping data. *J. Am. statist. Assoc.* **62,** 1159–1178.

GILMOUR, J. S. L. (1937). A taxonomic problem. *Nature, Lond.* **139,** 1040–1042.

GILMOUR, J. S. L. (1940). Taxonomy and philosophy. *In* "The New Systematics" (J. S. Huxley, ed.), pp. 461–474. Clarendon Press, Oxford.

GILMOUR, J. S. L. (1951). The development of taxonomic theory since 1851. *Nature, Lond.* **168,** 400–402.

GILMOUR, J. S. L. (1961). *In* "Taxonomy in Contemporary Botanical thought" (A. M. MacLeod and L. S. Cobley, eds), pp. 27–45. Oliver and Boyd, Edinburgh.

GOWER, J. C. (1974). Maximal predictive classification. *Biometrics* **30,** 643–654.

JOHN, S. (1970a). On identifying the population of origin of each observation in a mixture of observations from two normal populations. *Technometrics* **12,** 553–563.

JOHN, S. (1970b). On identifying the population of origin of each observation in a mixture of observations from two gamma populations. *Technometrics* **12,** 565–568.

PEARSON, K. (1894). Contributions to the mathematical theory of evolution. I. Dissection of frequency curves. *Phil. Trans. R. Soc.* A **185,** 71–110.

RAO, C. R. (1952). "Advanced Statistical Methods in Biometric Research". Wiley, New York.

SCOTT, A. J. and SYMONS, M. J. (1971). Clustering methods based on likelihood ratio criteria. *Biometrics* **27,** 387–397.

SNEATH, P. H. (1957). The application of computers to taxonomy. *J. gen. Microbiol.* **17,** 201–226.

SOKAL, R. R. and MICHENER, C. D. (1958). A statistical method for evaluating systematic relationships. *Univ. Kansas Sci. Bull.* **38,** 1409–1438.

WELCH, B. L. (1939). Note on discriminant functions. *Biometrika* **31,** 218–220.

WILLIAMS, W. T. and LAMBERT, J. M. (1959). Multivariate methods in plant ecology. I. *J. Ecol.* **47,** 83–101.

WOLFE, J. H. (1970). Pattern clustering by multivariate mixture analysis. *Multivariate Behavioural Res.* **5,** 329–350.

Teaching

16 | Computers in Some Instructional Aspects of Systematic Biology*

JOHN H. BEAMAN

*Department of Botany and Plant Pathology, Michigan State University,
East Lansing, Michigan, U.S.A.*

Abstract: Introduction of computers into undergraduate and graduate biology courses can provide the stimulus for students to apply the computer in their own work. A project at Michigan State University has been directed at development of classroom exercises using time-sharing computers in systematic biology, especially for identification of specimens. The exercises are based on a system of computer programs prepared by Larry E. Morse. They have been used in courses in introductory and advanced plant systematics, general biology, and a new course concerned specifically with computer applications in systematic biology. The programs have been implemented on a commercial international computer network, in which their effectiveness for specimen identification, key construction, description printing, and other applications has been demonstrated. Although it may be prohibitively expensive at this time to use a commercial system for research, instructional networks for systematic biology may now be economically feasible and could ultimately lead to a practical basis for processing and communicating computerized taxonomic research data.

Key Words and Phrases: classroom exercises, computer applications, data banking, identification, instruction, network, taxonomic data matrix, time-sharing computer

In his book "Man and the Computer", Kemeny (1972) stated that 90% of all students at Dartmouth College learn how to use a computer. He suggested that by 1980 no college or university should be given full accreditation unless computer services are freely available to all students. Although the Dartmouth example is exceptional, and universal availability of the computer to students within 7 years may be optimistic, the trend seems inevitable. Nevertheless, in

* Development of the computer programs and classroom exercises discussed here has been supported principally by National Science Foundation Grant GJ-573 and the Michigan State University Educational Development Program.

Systematics Association Special Volume No. 7, "Biological Identification with Computers", edited by R. J. Pankhurst, 1975, pp. 267–276. Academic Press, London and New York.

college and university instruction of biology, computers have received relatively little emphasis.

Computer applications in the educational process involve two very different approaches. On one hand they may serve as a direct means for increasing the efficiency or quality of education. Computer-aided instruction (CAI), computer-based education, and computer-managed instruction have this objective. Kemeny notes that CAI has come to stand for only one possible use of the computer—that of substitute teacher. For computer-based education in biology the development of both hardware and software has reached a fairly advanced stage in the PLATO system at the University of Illinois (Hyatt *et al.*, 1972).

Beyond its role as substitute teacher or administrator, a major category of involvement of the computer in education concerns its problem-solving usage. Principal applications in biology have been in statistical analysis, evolutionary and environmental modeling or simulation, and numerical taxonomy. Perhaps because these subjects are dealt with to a minor extent in introductory courses, little emphasis has been put on development of new classroom uses for computers in biology.

Many universities offer a rather broad array of computer science courses, including training in programming languages such as FORTRAN, ALGOL, and COBOL. However, few if any examples of the applications of computers to biological problems are likely to be included in such courses. Curriculum structure or inadequate motivation may keep many biology students out of computer science courses until they are well advanced in a graduate program and need the computer to analyze thesis data. These students are likely to be unaware of the full potential of the computer for their research, and their usage will be minimal. Many traditionally trained biologists probably avoid using computers simply because they never learned how to access the system. Introduction of computer applications into undergraduate and graduate biology courses can provide the stimulus for students to use the computer and should be a significant educational objective in the computer age.

A number of computer programs are now available to aid the systematic biologist, but these are highly diverse in both data-input formats and hardware requirements. As a result the average researcher must either submit his data to a specialist in a different institution or expend considerable effort to get the programs running on a locally available computer. These problems have tended to discourage widespread use of important programs and are undoubtedly factors retarding the introduction of computer methods in the classroom.

At Michigan State University (MSU) we have attempted to help improve this situation by developing classroom exercises which demonstrate how the computer may aid in the solution of certain problems in systematic biology. In introductory courses in systematic botany and general biology, we have provided "hands-on" experience with a time-sharing computer used to identify biological specimens. More advanced students in a graduate-level plant taxonomy course have had the opportunity to prepare and process their data using a library of standard programs. We have also developed a new course on computer applications in systematic biology for advanced undergraduate and graduate students. The instructional applications we have designed have not had the CAI objective of lower cost education, but they should result in better trained students and encourage more biologists to use the computer.

The MSU project has been centered around a number of programs written by Larry E. Morse. The Morse system has been developed as an integrated program package and data structure for processing taxon-oriented data. The development of these programs began in 1966; they were first used in a classroom situation in 1969; a preliminary program package was distributed in 1970; and a more advanced package has just been completed (Morse, 1974).

The programs of the new package accomplish the following tasks: specimen identification in a conversational mode, key construction, description printing, taxon–taxon comparisons, preparation of inverted descriptions (i.e. listing taxa which have particular characters in common), and production of punched-card field keys. In addition to these main data-processing programs, the system includes a supervisor program, a program for editing data files, and two programs for processing character lists. Although several of the programs can be used in either time-sharing or batch mode, the intent has been to develop a system for a national or international time-sharing network, thus obviating the need for local implementation.

The conversational specimen identification program has been the basis for our introducing an exercise on computer-assisted identification in the introductory plant systematics course. Time-sharing computers have therefore been essential to implementation of this exercise, but our entire instructional program has emphasized time sharing. The ability to access the computer in conversational mode from the classroom or laboratory brings a unity to the instructional process which is impossible with batch processing systems. Kemeny (1972) credits time sharing with enabling virtually all Dartmouth College students to learn to use the computer. The late James A. Peters was one of the most ardent spokesmen for the advantages of time-sharing computers in biology.

In the Morse system all programs operate with a standardized data file, the "taxonomic data matrix" described by Morse *et al.* (1971a). This data matrix format has provided an outstanding opportunity for students at MSU and several other institutions to prepare data for use with particular taxonomic groups in which they were interested. The matrix can be an extremely important element in teaching systematic biology because it is simple in concept but elegant in its comparative method for organizing diagnostic and descriptive data for a group of organisms. Most of the matrices prepared to date by our students are preliminary and experimental, but they have covered a wide range of taxa among the angiosperms, gymnosperms, pteridophytes, non-vascular plants, and even a few animals. Because of the tentative status of most of the matrices, the number of data elements recorded in them is perhaps not very meaningful, but we have on file 110 matrices treating a total of 1893 taxa (or their equivalents) and 2061 characters (or their equivalents). Most of the data were prepared by MSU students and faculty, but several matrices were constructed by Flora North America personnel at the Smithsonian Institution.

Three different approaches to classroom use of computers in the instruction of systematic biology will be described below. Implementation of two of these approaches required relatively little change in the conventional structure of the courses. The third approach has involved development of a new course. In addition to the opportunities for computer usage that these courses have provided, a number of our students have conducted independent research projects for academic credit. Most of these special projects have concerned data preparation and testing of the Morse programs, but several students have also written their own programs.

INTRODUCTORY PLANT SYSTEMATICS

This course is taught each spring quarter and has an enrolment ranging from 220 to 250 students. An exercise on computer-assisted identification has been used for 5 years, and this is included in the laboratory manual for the course (Beaman, 1971a). Subsequent to a demonstration in the laboratory on use of the computer, the students come individually to complete the exercise during evening hours. After studying the specimens to be identified, each student spends about 10–20 min at the computer terminal, identifying four specimens. We have mostly used Michigan species of maples (*Acer*) as the material to be identified, but other data matrices and specimens are available.

Each student is expected to write a brief critique on the exercise. Some of the critiques are highly imaginative, and a number of them have contained useful suggestions for improving the programs or other aspects of the exercise.

The exercise has provided for many students their first opportunity to use a computer and to see it applied to a biological problem. The critiques often indicate that the project helped convey some fundamental aspects of systematics, such as the fact that organisms are classified on the basis on numerous correlated characters rather than single characters. The concepts of character, character state, and variation in a group of closely related organisms are probably better illustrated in this exercise than in any other part of the course.

An exercise similar to this has been used once in a general biology course in connection with the study of algae. Student evaluations of this exercise indicated that it was effective in increasing their understanding of algal characteristics and also improved their understanding of the use of identification keys.

ANGIOSPERM TAXONOMY

This course includes a survey of major flowering plant families with emphasis on their morphology, distribution, and evolutionary trends. The course is given in alternate years and has an enrolment of 15–20 students. We have had only one opportunity to utilize the computer for a term project in the course, but this seems to have been an exceptionally successful effort. Each student had the responsibility for constructing and evaluating a taxonomic data matrix for families in one or two subclasses of angiosperms. The project was organized so that all subclasses would be treated by one or more students.

The students understood that their investigations might have some relationship to the preparation of data for the Flora North America Program (FNA), and that they could at least provide an example of one way to approach the task of computerizing data at the level of plant families for a major geographical area. It was expected that they might discover some of the pitfalls in the system and suggest improvements so that the method finally evolved for FNA would have the best possible chance for success. A detailed account of the results of the project is presented in FNA Report 56 (Furlow and Beaman, 1971).

Most of the problems encountered by the class concerned neither the computer nor the programs but rather character selection and data collection, organization, and coding. Program and data format difficulties which were noted during the course have mostly been resolved. As a learning exercise the class was unanimously enthusiastic about the project. Not only did it reinforce facts and principles studied in class, but it also gave the students an opportunity to become familiar with the computer as a tool which interested them, but which to many had previously seemed untouchable.

COMPUTERS IN SYSTEMATIC BIOLOGY

This course was taught for the first time in the spring of 1970 and was again given at MSU in the winter of 1971. It was offered with slightly different emphasis under the title "Computadoras en Botánica" at the National Autonomous University of Mexico (UNAM) in April, 1972. Its enrolment of 38 students in the three classes included advanced undergraduates and graduate students. A large number of faculty and student visitors also participated. The course is described by Furlow *et al.* (1971), and a syllabus is the subject of FNA Report 62 (Morse *et al.*, 1971b). Six student term papers from the course, with examples of diverse uses of the taxonomic data matrix, are given in FNA Report 63 (Beaman, 1971b).

A major aim of the course is to provide students with a perspective of current and projected computer applications in systematics and related fields. Topics discussed in the MSU course are as follows: a general consideration of computers in systematics; the Flora North America Program; taxonomic data bank concepts; specimen identification and medical diagnosis; plant disease diagnosis by computer; information problems in paleontology; distribution mapping by computers; biometrics and statistical studies with computers; probabilistic approaches to specimen identification; computers as aids to teaching systematic biology; design and development of instructional systems; numerical taxonomy; modeling of genetic and environmental systems; possible computerized nomenclatural aids; quantitative phyletics; and biogeographical analysis with computers.

This type of course should stimulate development of new uses for computers in systematic biology, and one of the best ways to encourage this type of student creativity is through individual class projects. The taxonomic data matrix was used by a number of students for preparing data on various angiosperm taxa, but others applied it to pteridophytes, basidiomycetes, potato diseases, and pollen. Extensions of the matrix method were used for analyzing species composition in a local flora, selection of ornamental plants, recording and retrieving phenological data, analyzing cereal leaf beetle behavior, and diagnosing psychological disorders. Methods of numerical taxonomy and use of biometric procedures were the subject of several student reports, and other topics explored include evaluating parameters of population dynamics, information systems analysis, and development of an information system for insect and plant relationships.

Although it is helpful for students enrolling in this type of course to have already had computer programming experience, such background is not entirely necessary. If a library of programs is available, the student without programming experience can use these for processing data which he prepares. The Morse programs have been especially useful for this purpose. In a number of instances the initial experience with computers gained by students in this course has encouraged them to enrol in computer science courses and utilize the computer in thesis-oriented research.

For the past 3 years the Morse program system has been operational on an experimental basis on the General Electric Mark II Network. This can be accessed with a local telephone call from more than 250 cities in North America, western Europe, and Japan. In addition to use of the programs and data files at MSU, they have been accessed through the network at the Smithsonian Institution and Harvard University, where some of the programming has been done. The network was used to demonstrate the Morse programs at FNA workshops during scientific meetings at Indiana University and the University of Alberta. By means of the network the programs and data files were used in the Smithsonian Summer Institute in Systematics in 1970. Graduate students at MSU, Duke University, Rutgers University, and the University of Texas have used the network and programs for their thesis data. Scientists in the taxonomy laboratory of the Agricultural Research Service, U.S. Department of Agriculture in Beltsville, Maryland, have used the network and programs for developing a data bank for seed identification. The system has been demonstrated via the network for botanists at the National Museum of Natural Sciences, Ottawa, Canada.

The versatility of the Morse programs for network implementation is illustrated by the fact that earlier versions were operational for about a year on two other commercial networks besides the GE Mark II. The program for direct identification has also been adapted by A. L. Jones of MSU (Jones and Harsh, 1970) for identification of plant diseases. This program has been implemented on the University of Michigan computer in voice mode with touch-tone telephone terminals for an experimental network used by county agents throughout Michigan.

The network approach to data banking has already become well established in higher education for bibliographic purposes. For example, the MSU library currently has on-line access to data on medicine, agriculture, chemistry, and educational resources through subscriptions to the MEDLINE, CAIN, CHEMCON, and ERIC networks. Another example of network availability and potential is illustrated by the MERIT network between MSU, the University of Michigan and Wayne State University. The computers at these institutions are available for classroom and research use on a virtually equal basis for persons at any one of the institutions. The Morse system could be readily implemented on this network.

The major barrier at present to a network for instruction and research in systematic biology is economic. An hour of real time on the GE Mark II can amount to \$20–50, depending upon the time of day and nature of programs and data files used. In addition there are minimum monthly charges and computer storage costs. In contrast to these commercial costs, some of the Morse programs and data files have been run on the MSU CDC 6500 time-sharing system for \$3–4 per hour. For classroom usage only 12% of this amount is assessed to the course.

Another problem in implementing a network for research grade taxonomic data is that the data do not now exist in machine-readable form, and are not likely to become available in the near future (Shetler, 1974). User demand could not at present make a research network self-sustaining. On the other hand, a relatively small data base could support an instructional network, and programs such as those in the Morse system could be made widely available for classroom purposes. The initial implementation of a primarily instructional network might ultimately result in the development of a sufficiently large data base and low enough costs to permit research applications on the same system. As research data were generated, these would also be available for instructional uses.

The taxonomic data matrix in the Morse system is particularly well suited to instructional network implementation. Through use of chained subordinate matrices in the data file hierarchy, such as matrices for families of an order, genera of a family, and species of a genus, a large data bank can be accommodated, and it could be built step by step. The potential access to a large, hierarchical data base is an unusual if not unique feature of this instructional application of computers. Special purpose matrices for various teaching functions could be designed, and a hierarchy of relatively small data sets would facilitate processing on small computers when necessary.

Network implementation of programs useful in systematics would make it possible for instructors with little or no programming experience to use the programs and data bank for their classes. A network would also have a significant advantage over local implementations in saving on computer storage costs, since one copy of each program and data file would serve all users and the costs could be shared among the entire user community. Additional systems advantages of the matrix in a network environment are discussed by Morse (1974).

Although a full-scale time-sharing data network such as could be implemented on the GE Mark II would be optimal for many data processing needs in systematic biology, both teachers and researchers may have to accept a less sophisticated and certainly less expensive system for the present. The informal kind of information, program, and data exchange advocated by Peters (1972) can be continued and expanded. Programs such as those developed by Morse can be implemented on local time-sharing computers, and some of his programs can even be operated in the batch mode although several important options are thereby eliminated. Some systematists, especially those in smaller institutions, may be able to obtain free access to under-utilized computers.

CONCLUSIONS

In the instructional milieu of plant systematics at Michigan State University the computer has served at the very least as a clever device to interest beginning students in both systematics and computers. For more advanced students at MSU, the University of Mexico, and several other institutions, the Morse programs have opened a vista to the potential of the computer for analysis and retrieval of biological data. The system has provided an important method for processing taxonomic data in the thesis research of several students. Both the hardware and software are now available for expanding the opportunities for biological applications of computers. The limiting factors are economics and acceptance by biologists.

Development of the Morse system has been more or less concurrent with that of the Flora North America Program. Computer-constructed keys and diagnostic descriptions from the Morse programs provided a dramatic illustration for the FNA Editorial Committee in 1969 of the power of the computer to process taxonomic data. This demonstration was critical to the decision to make FNA a computerized data bank. The MSU research on computer-assisted identification has subsequently been one of the FNA pilot projects on automated data banking.

With the entire FNA effort now suspended (Walsh, 1973) and its future prospects gravely jeopardized, the Morse system may offer one of the best opportunities for continuing development of an international botanical or biological data bank. Similar programs, such as those by Dallwitz (1974), Gower and Barnett (1971), Hall (1973 and this volume), Pankhurst (1971 and this volume), and Payne (this volume) likewise offer promise. Continuation of the FNA Program would depend on the reactivation of a large and highly organized staff and access to the powerful and expensive IBM Generalized Information System. FNA cannot be operated as a scaled-down effort. Computer programs such as the Morse system, however, can be implemented as modest instructional networks and on local time-sharing computers as well. They may serve effectively in capturing the interest of biology students in electronic data processing and helping train them in data banking techniques.

In the absence of FNA such programs still offer hope for building the major biological information system of the future.

REFERENCES

BEAMAN, J. H. (1971a). "Introductory Plant Systematics: Lecture Syllabus and Laboratory Manual". (4th edition) Department of Botany and Plant Pathology, Michigan State University, East Lansing.

BEAMAN, J. H., ed. (1971b). Some applications for the taxonomic data matrix: six term papers by students at Michigan State University. *Flora N. Am. Rep.* **63**, 1–103.

DALLWITZ, M. J. (1974). A flexible computer program for generating identification keys. *Systematic Zoology* **23**(1), 50–57.

FURLOW, J. J. and BEAMAN, J. H. (1971). Sample taxonomic data matrices for vascular plants prepared by students at Michigan State University. *Flora N. Am. Rep.* **56**, 1–118.

FURLOW, J. J., MORSE, L. E. and BEAMAN, J. H. (1971). Computers in biological systematics, a new university course. *Taxon* **20**, 283–290.

GOWER, J. C. and BARNETT, J. A. (1971). Selecting tests in diagnostic keys with unknown responses. *Nature, Lond.* **232**(5311), 491–493.

HALL, A. V. (1973). The use of a computer-based system of aids for classification. *Contr. Bolus Herb.* **6**, 1–110.

HYATT, G. W., EADES, D. C. and TENCZAR, P. (1972). Computer-based education in biology. *BioScience* **22**, 401–409.

JONES, A. L. and HARSH, S. B. (1970). Automated consulting system developed. *Phytopath. News* **4**(10), 3.

KEMENY, J. G. (1972). "Man and the Computer". Charles Scribner's Sons, New York.

MORSE, L. E. (1974). Computer programs for specimen identification, key construction, and description printing using taxonomic data matrices. *Publ. Mus. Michigan State Univ., Biol. Ser.* **5**, 1–128.

MORSE, L. E., PETERS, J. A. and HAMEL, P. B. (1971a). A general data format for summarizing taxonomic information. *BioScience* **21**, 174–180, 186.

MORSE, L. E., FURLOW, J. J. and BEAMAN, J. H. (1971b). Computers in systematic biology: a course syllabus. *Flora N. Am. Rep.* **62**, 1–58.

PANKHURST, R. J. (1971). Botanical keys generated by computer *Watsonia* **8**, 357–368.

PETERS, J. A. (1972). Museum and university data, program, and information exchange. *MUDPIE* [Mus. Univ. Data Prog. Info. Exch., Smithsonian Inst.] no. 26, 2–6.

SHETLER, S. G. (1974). Demythologizing biological data banking. *Taxon* **23**, 71–100.

WALSH, J. (1973). Flora North America: Project nipped in the bud. *Science* **179**, 778.

DISCUSSION

The Future of Automatic Identification Methods for Higher Plants*

A. GÓMEZ-POMPA

Department of Botany, Institute of Biology,
National University of Mexico

BACKGROUND

In order to foresee the future of any new technique we have to keep in mind the possible users and the advantages it has over other techniques.

In the case of identification of higher plants the techniques available are: the keys published in Floras, Revisions, Monographs and other similar written sources; or the "matching method", when a herbarium is available and the higher taxa (family and genus) are known; the third possibility is consultation with expert plant taxonomists. There are other procedures to identify plants (local names, photos, checklists, etc.) but they should be considered as variations of those already mentioned.

The users of the techniques available for identification of higher plants may be divided in three groups. The first is the layman who is interested in learning the name of a given plant for practical reasons, e.g. in the case of a poisonous plant or a wild flower for his garden. The second is the professional person who must know the name of the plant for his own work (e.g. an agronomist who needs to identify a weed; or a lawyer a narcotic plant, etc.). The third is the professional botanist who studies plants and must know their names; in this case I am referring not to plant taxonomists, but to other botanical scientists.

All these users will have to follow any of the identification techniques available. It is quite clear to me that the techniques available have proved to be very effective only for those areas where the necessary publications are available,

---

* Flora of Veracruz, contribution number 16. A joint project of the Institute of Biology of National University of Mexico and the Department of Botany of the Field Museum of Natural History to prepare an ecological floristic study of the State of Veracruz, Mexico. Information about this project was published in *Anal. Inst. Biol. Univ. Nacl Mex. Ser. Bot.* **41,** 1–2. Partially supported by N.S.F. Grant GB-20267X.

where herbaria exist and are available and when the time of the existing experts
for this purpose is not a limiting factor.

The main problem emerges when these techniques or facilities are not
available. This may be the case in most parts of the world, where floras need to
be described, where monographs and revisions are lacking, and where local
taxonomic expertise is nil or heavily involved in other projects. Another
problem we have to solve is the need for faster and more efficient methods of
identification. It is obvious to many of us that this is a great problem and
deserves more attention in the near future. The number of precise identifications
of plants is increasing very rapidly. In Mexico last year we had more
consultations at the National Herbarium than in any previous year in its history;
the same can be said of the year before last. Many land-use projects, ecological
studies, environmental impact statements etc. need help with plant identification
but our capabilities are far behind their requests.

I do not know how general our case is, but from informal conversation with
other colleagues it seems that it is a widespread situation.

For this reason it is obvious to me and my colleagues in Mexico that one top
priority is the production of means of identification (Floras, Revisions,
Monographs, checklists) as quickly as possible to discharge the heavy consultation
load on the few individual taxonomists.

Close attention has to be given to the areas or groups of plants of economic
importance, or regions that are in danger of destruction with the consequent
extinction of biotypes or species.

A topic of great concern is the education of new botanists who will help to
bring closer the solution of the many problems concerning the knowledge of
our Flora.

For these and other reasons (Gómez-Pompa and Nevling, 1973) a computerized Flora Program of the State of Veracruz has been started to explore
new, faster, and more efficient ways of producing a Flora. During the development of the program we have considered, on several occasions, the possibility
of developing automatic identification facilities within our computer
system.

AUTOMATIC IDENTIFICATION TECHNIQUES FOR HIGHER PLANTS

It is clear that the problem does not lie in the software for computer identification:
there are programs and packages of programs available for this purpose (see
chapters by Morse and Pankhurst, this volume) that have been tested on many
different groups and no major problem seems to appear. It is also clear that these
procedures will evolve to produce more efficiency, as it has been demonstrated

by Pankhurst (1972) in his simple but sophisticated method of data capture. It is also important to mention that hardware is not an obstacle, as every day these facilities are more and more widespread and their costs, through time-sharing, are of lesser importance.

For all these reasons we have been watching the development of the field of automatic identification for the Flora of Veracruz Program. At the same time we are testing these methods to implement future specific projects. It has been decided that our program will have some pilot projects in this line, setting some priorities according to the demand for identification. What seems more useful to our program is the use of automatic methods as an aid to key construction.

I believe that automatic identification methods for higher plants will play a secondary role, on a worldwide basis, as a general scheme for identification of plants. They may play a fundamental role in the process of speeding up the development of Floras by contributing to key construction. They should also play a fundamental role in routine identification of taxonomic groups of economic importance and of local plants of special interest (e.g. those in experimental stations, biological reserves, etc.).

The most important restriction for general use of these methods will be the amount of time available for data gathering and input into a computer system. The data may be obtained from local Floras (or monographs, etc.) which may turn out to be a duplication of effort because they already have built-in identification techniques. Furthermore, and if these publications are not available, in the process of description of a new Flora (or Monograph, etc.) for automatic identification, it may be more generally useful if the end result is a written product with identification keys; although, in this last case, it may be advisable to have the new descriptions in such a form that they can be used in automatic identification. This could be done through a computer terminal or by means of facilities to reproduce outputs to help in the identification procedures, such as the production of card decks (polyclaves) for field use, e.g. the sample shown by Pankhurst (this volume).

On the other hand, the lack of computer facilities may be a great handicap for the implementation of these methods. For this reason the techniques may be confined in the future to large institutions having heavy public demand for identifications.

In spite of all these problems it seems to me that the future of this field in the identification of higher plants will be of primary importance because of the great need for documentation of the diversity of plants in their natural environment that humankind is destroying at an alarming rate. I cannot predict the future of these methods, but I am sure that they already have their place in the biological sciences, and that we have in our hands a valuable tool to be used intelligently.

REFERENCES

GÓMEZ-POMPA, A. and NEVLING, L. I. Jr. (1973). The use of electronic data processing methods in the flora of Veracruz program. *Contr. Gray Herb.* **203,** 49–64.
PANKHURST, R. J. (1972). A method for data capture. *Taxon* **21**(5/6), 549–558.

A Botanist's View of Automatic Identification

J. McNEILL

*Biosystematics Research Institute, Research Branch, Agriculture Canada,
Ottawa, Canada*

I feel that it is appropriate to look from a strictly biological standpoint at the value and applicability to systematics of these computer-based techniques. This I shall try to do as a vascular plant taxonomist and therefore I must include the *caveat* that my thoughts are based on the nature and stage of development of higher plant systematics. Although I am sure that most of what follows is applicable to many other groups, particularly the morphologically complex, there are certainly special considerations that apply to some, notably the various groups of micro-organisms.

UTILITY OF AUTOMATIC IDENTIFICATION

As this symposium has amply demonstrated, it is when identification by more traditional methods proves difficult that biologists turn to automatic techniques. This is as it should be, because what is important is not the method of identification but its accuracy. The least costly method (in terms of effort, time or money) that provides the correct name is evidently the best. So if more elaborate methods are proposed because of difficulties with existing ones, we should ask what the reasons are for identification proving difficult. I suggest that there are four major causes of difficulty.

1. Inadequate Basis for Comparison

Identification is, as Walters (this volume, Chapter 1) and others have already pointed out, the assignment of a specimen to a particular position in a pre-existing classification. So whatever method is used, a comparison is made between the specimen and some expression of the data-base of the classification, whether this be a printed key or description, an electronic data-bank, a folder of "authentic" herbarium specimens, or even the mental image of a species possessed by a knowledgeable systematist. It is convenient to recognize two

K   283

kinds of inadequacy, although as we will see there is a sort of inverse relationship between them.

(a) *Inadequate material*. The specimen lacks organs necessary for identification in that group (e.g. an umbellifer without fruit) or is altogether too fragmentary (e.g. a single chewed-up leaf).

(b) *Inadequate reference base*. The classification by which the identification has to be made is badly presented. Taxonomists, even when they understand their taxa thoroughly, are unfortunately not always able to present diagnostic features clearly or even correctly, and we all know that there are monographs and Floras in which the keys never seems to "work" very well, or in which the most reliable diagnostic features are not included. In other cases it may just be that the published treatment requires more data for identification than seems reasonable (e.g. in a dioecious group, characters from both the male and the female). With a perfect specimen identification may be possible even if the classification is poorly presented but the more incomplete the specimen, the better the reference base must be.

2. Inadequate Taxonomy
Besides the technical inadequacies described, a specimen may not be identifiable in terms of a particular classification for two other reasons, which are more specifically biological.

(a) *Taxon not included*. The specimen may represent a species or other taxon not included in the existing classification, because it is either a new record for the area or in fact completely new to science. If, on the other hand, the specimen represents an unreported hybrid between two known taxa or even an unusual growth phase of one, we approach the next problem situation.

(b) *Taxa are unsatisfactory*. The existing classification may simply be bad, in the sense that at least some of the taxa recognized do not represent biological populations in any meaningful way. In the only useful sense of the word, the taxa are "unreal" and unless they are crudely monothetic, satisfactory assignment of specimens to them is unlikely. The same problem of unsatisfactory taxa may exist less because of taxonomists' failures than because of the particular reproductive and evolutionary mechanisms operating (e.g. agamospermy, introgressive hybridization, abrupt speciation etc.). In such cases it may be impossible to summarize, usefully, the variation patterns existing in nature in a hierarchic classification with the species as its unit. From my experience in vascular plant taxonomy it is in these situations, in which the existing classification is unsatisfactory, that the major problems in specimen identification rest.

3. *Too Many Taxa, So Much Alike*

Here it is the sheer size of the problem that presents the major difficulty. Even if the question of facultative agamospermy did not arise in the recognition of *Rubus* species, Pankhurst's paper (this volume, Chapter 14) and demonstration showed that the number of taxa and characters involved is so great as to make accurate identification by any method extremely difficult, and by conventional keys almost impossible. This is true of any situation in which there is a substantial number of taxa broadly similar to one another and differing for the most part in characters which the human eye interprets as insignificant.

4. *Two Few Taxonomists*

Of course the lack of good modern monographic taxonomy lies behind the first two difficulties that I have discussed, but under this heading, I am considering specifically the current lack of personnel sufficiently well trained biologically and taxonomically to use successfully the full battery of conventional identification tools—manuals and Floras, revisions and monographs, keys and checklists, reference sets and the major museum and herbarium collections. In institutes responsible for taxonomic identification services, this problem is recognized as pressing, particularly because of the recent and continuing increase in the demand for identifications with increased concern about the quality and composition of man's environment.

There may be other factors, but I believe that these four headings—Inadequate basis for comparison; Inadequate taxonomy; Too many taxa, so much alike; and Too few taxonomists—summarize the main reasons for the difficulties which exist in biological specimen identification, and as we have seen it is when difficulties exist that we turn from conventional methods to new ones such as computer-assisted techniques. I would venture to suggest, however, that it is only for the problems encompassed by headings 3 and, in certain cases, 4 that automatic methods are in any sense a solution. What are, in my experience, the most widespread causes of difficulty in identification are represented in headings 1 and 2. These are essentially failures at the primary classification phase rather than at the secondary identification one, and the inability of any change in identification technique to solve these problems must be emphasized. To ignore this fact would be to do a disservice to automatic identification. I say this because I believe that computer-based techniques have an increasingly important role to play in those limited problem areas for which they are appropriate.

CHOICE OF AUTOMATIC IDENTIFICATION METHODS

Having set up a unidimensional classification of identification problems, it is natural in a symposium such as this to think in matrix terms, and now consider different automatic identification methods in the light of their suitability in problem areas of vascular plant identification. The "classification" of methods of computer-assisted specimen identification presented earlier by Morse is a good one upon which to base this comparison. As seen from Table I (p. 13), he recognizes eight main methods, some of which represent monothetic and polythetic versions of the same basic technique. Let us look at these methods, not in Morse's order, but in what I believe to be a sequence of increasing usefulness in the taxonomy of morphologically complex organisms such as the vascular plants.

1 and 2. On-line Polyclave (Monothetic or Polythetic)

As anyone who participated in the Flora North America demonstrations over the past few years will attest, the use of this type of identification tool is a fascinating and exciting experience. However, the requirement of on-line storage of the data-base makes its practical utility extremely low, and applications are likely to be restricted to very special situations where it is economically important to have an immediate identification as between a relatively limited number of possibilities (e.g. the highly poisonous species of a particular area, the other species being treated for this purpose as a single taxon "Not-poisonous"). As a teaching tool, however, its potential for interesting and stimulating students is unquestionably great (cf. Beaman, 1971).

3 and 4. Character-set Matching (Monothetic or Polythetic)

As Morse points out, only the polythetic case need be seriously considered as a computer-assisted method. Despite its apparent success in various groups of micro-organisms (cf. e.g. Gyllenberg and Niemelä, Chapter 9), I am very sceptical about its general utility in vascular plant identification, in which character-states are more conveniently recognized one by one whereas the results of a set of tests on cultures of micro-organisms, for example, are more conveniently recognized in batches. I suspect that the relative advantages which Pankhurst (Chapter 14) demonstrates arise from the poor quality of the *Rubus* data-base which he was required to use, and it may be that any usefulness of this approach in vascular plant identification is in helping to resolve type 1 difficulties, for which the use of automatic methods is otherwise not advantageous.

5. *Computer-generated Key* and 6. *Card-key Generator*

These are the monothetic and polythetic analogues of each other, and I would like to consider them together, because I believe that the greatest contribution of computer-assisted identification to taxonomy lies in these two groups of methods. When dealing with relatively small numbers (perhaps less than 15) of rather similar taxa over which he has worked for some time, a taxonomist can usually produce a good key directly, in a shorter time than it would take to produce a data matrix for a computer-generated key or card-key. However, as soon as the number of taxa increases, or they become more diverse, or less intimately known by the taxonomist, the value of computer-assistance is striking. Pankhurst has demonstrated this by producing a key to the genera of European Umbelliferae which although derived from Professor T. G. Tutin's data base, had about 20% fewer leads, used fewer characters and was line by line more compact than that prepared by Tutin for Volume 2 of Flora Europaea (cf. Pankhurst and Walters, 1971). Other examples have been given in this symposium. The choice between a computer-generated key and a card-key deck depends partly on technical requirements—only the former is suited to a printed monograph or Flora—and partly on the importance of the multi-entry (polythetic) capability of the card-key. The latter is obviously required in any situation in which incomplete specimens are frequent; this is equally true whether the card-keys are computer- or manually generated (cf. the early use of manually generated card-keys for timber identification by Clarke, 1938, and Phillips, 1941, 1948).

Two of Morse's identification methods have yet to be mentioned. The first is the computer-stored key which seems to have little practical utility and the second, *Automatic pattern recognition* procedures. This "super-polythetic" approach is, of course, very much in its infancy and as Morse has said we still do not know how successful future developments will be. However, at the risk of seeming unimaginative, I must say that I find it hard to envisage any general application in vascular plant taxonomy, largely because of the indeterminate growth of plants. There are, however, plant structures with more determinate growth characteristics and where there is an economic need for screening and identification of very large numbers of these, automatic pattern recognition may prove valuable. One possible application is in fossil pollen grain recognition, but as with other on-line methods, only if identification to one of a limited number of taxa were adequate. The rest of the enormous diversity of grains would then be in a residual "other grains" taxon.

In conclusion let me first of all emphasize again what I believe to be the limited role of any developments in identification technique. If existing

K2

classifications are unsatisfactory, more effective results will be achieved by taxonomists revising the taxonomy of the groups concerned than by using more powerful tools on the existing inadequate structure. This should not be regarded as a dampener on the proceedings of this meeting, but rather as an attempt to ensure that a very valuable new methodology will not be discredited by inappropriate applications.

Although the utility of automatic identification is dependent on good classification, there is a contribution that the development of automatic methods can make to classification. This is the salutary effect that the discipline of preparing an accurate data-matrix has on the preparation and presentation of the classification. Pankhurst (Chapter 14) has already referred to the inconsistencies in the *Rubus* monograph (Watson, 1958) and I have found the same difficulties in attempting to use published data for numerical taxonomic work. I hope that one of the by-products of the use of the techniques we have been discussing will be greater precision in the description of taxa and in the presentation of taxonomic data.

Finally, what of the future development of automatic methods? There are still limitations in the extent to which variable characters, particularly quantitative ones, can be used in key construction. Perhaps the utilization of Jardine and Sibson's (1970) suggestions for more precise specifications of ranges will permit more flexible use of such characters.

Neither Morse nor Pankhurst may feel disposed to accept my strictures on the limited applicability of the techniques which they have developed, and I look forward to a spirited discussion. I am certain, however, that there need be no debate as to whether or not automatic methods of identification have a place in biological taxonomy. They have a most important one and so far as the higher plants are concerned, I have tried to define what that is.

REFERENCES

BEAMAN, J. H., ed. (1971). Some applications for the taxonomic data matrix. *Flora N. Am. Rep.* **63**. Washington, D.C.

CLARKE, S. H. (1938). A multiple-entry perforated-card key, with special reference to the identification of hardwoods. *New Phytol.* **37**, 369–374.

JARDINE, N. and SIBSON, R. (1970). Quantitative attributes in taxonomic descriptions. *Taxon* **19**, 862–870.

PANKHURST, R. J. and WALTERS, S. M. (1971). Generation of keys by computer. *In* "Data Processing in Biology and Geology" (J. Cutbill, ed.), pp. 189–203. Academic Press, London and New York.

PHILLIPS, E. W. J. (1941). The identification of coniferous woods by their microscopic structures. *J. Linn. Soc. (Bot.)* **52**, 259–320.

PHILLIPS, E. W. J. (1948). Identification of softwoods by their microscopic structure. *Forest Prod. Res. Bull.* **22**, H.M.S.O., London.

WATSON, W. C. R. (1958). "Handbook of the Rubi of Great Britain and Ireland". Cambridge University Press.

Classification and Identification

W. W. MOSS

Academy of Natural Sciences, Philadelphia, Pennsylvania, U.S.A.

IMPLICATIONS OF THE CLASSIFICATION/IDENTIFICATION PROBLEM

As Gower has noted (Chapter 15), a considerable confusion between the terms classification and identification has existed in the literature of taxonomy until recently. The currently recognized distinction between classification (the recognition of groups) and identification (the assignment of unknowns to preexisting groups) may not be, however, quite as simple as we had thought. It seems reasonable to suggest that we classify to recognize order in the apparent chaos that surrounds us, to determine patterns of phenetic similarity, and (indirectly) to estimate ancestral relationships. Gower's intriguing suggestion is that in fact we classify, consciously or unconsciously, for quite another reason altogether, and that is to optimize identification. This is certainly food for thought, and points to a need for further analysis of what might be called the "motivational" or "psychological" aspect of taxonomic activity. Studies analysing the aims, motivations and operations of taxonomists have been curiously rare to date, possibly because of the "gentlemen's agreement" (Moss and Hendrickson, 1973) that permeates taxonomy; the tendency to assume that an authority knows what he's doing simply because he is an authority can lead to problems!

The distinguished historian of science, Loren Eisely, commented (1961) on the implications of the fact that western science arose from a civilization noteworthy for the intensity of its faith in a divinely impressed order. This faith served for centuries to stifle scientific thought by discouraging the asking of relevant questions. Ironically, however, faith also came to benefit the progress of science, for scientists tend to approach the resolution of problems in a logical manner, strengthened by the assumption that some underlying, logical and analyzable structure is in fact there and waiting to be discovered. Some experimental evidence exists to show that we as taxonomists look for groups and gaps because we firmly believe that they are there, and that we tend to see them whether they are there or not (Moss, 1971). The recent work of Hansell and Chant (1973) has

290

shown how firmly the dead hand of the past lies over subsequently recognized classifications; and Walters in Chapter 1 has noted clearly the difficulty in distinguishing between conservatism and stagnation during 300 years of stasis in the carrot family (Umbelliferae). Finally, Phipps' recent (1972) studies of the taxometrics of changing classifications include his quantitative evaluations of the conflicting results of different workers. Studies such as these suggest that a similar approach might well be used to investigate the utility of various automated identification schemes. For example, studies might be made of user reactions to such schemes, with an attempt to determine just what is the best approach to follow for rapid and useful identification. We have traditionally used a printed, dichotomous key, but might this be more because of its value to the monographer than to the user? All of us are familiar with traditional keys that tend to chain one to what often seems to be an arbitrary sequence of characters which may or may not be obtainable from the specimen to be identified. And almost all of us have attempted to identify a single specimen with a key that started out with characters of the female and ended with those of the male! The free, interactive input of a variable number of character states in no particular order (as in the keys of Pankhurst and Aitchison, and Morse) certainly seems a much more desirable approach.

The overwhelming majority of quantitatively oriented work in taxonomy in recent years has been addressed to classification rather than to identification. I cannot help but feel that this is due at least in part to a belief among numerical workers that significantly more intellectual challenge lies in the recognition of classifications than in the subsequent devising of schemes of identification. In this view, automated identification comprises basically a translation of orthodox keys into computerese, a clean-up job for the camp followers of classification. Regrettably, some of the early attempts to produce automated keys have not tended to allay this suspicion. In contrast, it seems clear from the present conference that some of the more recent developments in automated identification are beginning to provide more sophisticated options than have been available in orthodox identification schemes. In addition, the devisers of automated keys are beginning to encounter problems in both logic and methodology that might not have been predicted in advance. Glancing backward, much of the early work in numerical taxonomy is now considered to be somewhat naive, although for its time it had a heuristically balanced element of technological sophistication and intellectual challenge. Automated identification seems to be going through a similar, early period of excitement. As Pankhurst has noted, however, it seems wise to guard against an overemphasis on technology in the absence of a sound basis in theory.

THE PURPOSES OF IDENTIFICATION SCHEMES

It seems to me that more attention might be paid to the question: what are the purposes of identification schemes? Identification schemes might be set up in order to satisfy at least two different criteria: first, to identify an unknown in a way that somehow reflects its classification; and second, and more pragmatically, to obtain an answer quickly using the system that is most acceptable to the user (or desirable to the deviser). An example of the first might be a key that provided the most probable identification (species "X") accompanied by a list of other species within the genus (or at least judged to be close relatives in the most recent revision). A key of this nature would require continual modification with changing taxonomic opinions. An example of the second kind of key might be a scheme that provided the most probable answer accompanied by a list of other taxa that seemed to be a close fit solely on the basis of the data input. It would seem to me that the latter is the more desirable approach for practical purposes, and it seems to be the approach most popular here.

Interestingly, the "phenetic–phylogenetic" controversy so familiar to numerical taxonomists does not seem to have been a significant bone of contention in automated identification. This is possibly due to the early recognition by key-makers (as noted by Pankhurst) that phylogenetic keys are merely identification schemes based on restricted data sets, a concensus that should have been reached much earlier for classification, and which would have settled much of the conflict that still surrounds the role of phenetics and phyletics in classification today. The problems inherent in the construction of a phylo-genetic key did not always deter earlier workers, however. An example of a rather intriguing phylogenetic key that springs to mind is one that was once produced for a group of mites (J. H. Camin, personal communication). The key separated two supraspecific taxa on the condition of their epigynial shield, as follows:

> Epigynial shield primitively absent ....Taxon X
> Epigynial shield secondarily lost ....Taxon Y.

Clearly such keys have an explanatory potential, but they are of limited use, except perhaps to the deviser (or an extremely long-lived user!).

SOME PRACTICAL CONSIDERATIONS

On the basis of the papers presented here over the past two days it would seem that we now have a choice of several worthwhile systems; unfortunately, in some cases these systems also come liberally provided with their own sweeping promises and advertisements. One of the distinguishing features of numerical taxonomy has been the appearance of a variety of program packages for classification, each enthusiastically championed by the person who constructed it; this is a pattern which seems slated to be repeated for automated identification. Presumably the value of all such schemes will be decided through use and time, and hopefully we shall be spared the high-pressure salesmanship that can develop from overly unbridled enthusiasm.

It is obvious from the presentations here that some further form of character analysis is going to be necessary in automated identification as well as in automated classification. The basic questions are, of course, how many characters should we include, how many should we delete and what should be the criteria for such inclusion or deletion? Schemes of character analysis being investigated currently in numerical classification should be of interest to the producers of automated keys.

Automated data recording as described by Pankhurst is an intriguing way around what is without doubt the most dull and unappealing aspect of taxonomy. Obviously, his scheme would be best applied to a large group where the characters have previously been well defined and analysed, and for which subsequent entries of new species are frequently made. I would imagine that the preliminary programming requirements would make this approach somewhat inefficient for taxonomic revisions of small groups in which characters and character states had to be redefined frequently.

On-line key interaction is a particularly attractive teaching tool, and is useful as a demonstration of computer potentialities for automated identification. It may not be practical as a tool for the professional, or in situations outside of class, although this judgement would clearly be disputed by some. There are obvious places where the usage of available on-line keys could be speeded up and improved (for example, the laborious keyboard input of a character description could be avoided by simply entering the character number and having the machine echo the character description as a check).

Finally, it would seem useful for a practicing taxonomist to have some linkage of automated classificatory and identificatory programs, now that both types of programs are becoming freely available.

To return once again to the dichotomy between classification and identification, it is perhaps worth noting that there exists today an organization, The Classification Society, which concerns itself primarily with principles and methods of automated classification. It would seem wise for this Society to consider expansion of its horizons to include automated identification, or this and similar conferences may serve to mark the beginnings of an Identification Society!

REFERENCES

EISELEY, L. (1961). "Darwin's Century". Doubleday Anchor, Garden City, N.Y.

HANSELL, R. I. C. and CHANT, D. A. (1973). A method for estimating relative weights applied to characters by classical taxonomists. *Syst. Zool.* **22**, 46–49.

MOSS, W. W. (1971). Taxonomic repeatability: an experimental approach. *Syst. Zool.* **20**, 309–330.

MOSS, W. W. and HENDRICKSON, J. A. Jr. (1973). Numerical taxonomy. *A. Rev. Ent.* **18**, 227–258.

PHIPPS, J. B. (1972). Studies in the Arundinelleae (Gramineae). XI. Taximetrics of changing classifications. *Can. J. Bot.* **30**, 787–802.

Comments on the Automatic Identification of Biological Specimens

CHRISTOPHER WILKINSON

Department of Biological Sciences, Portsmouth Polytechnic, England

Although some senior biologists regard computers in biology, particularly in taxonomy, as an irrelevance, the number of questions arising from Beaman's paper (Chapter 16), (and one the writer gave (1973b)), shows that there is considerable opinion in favour of teaching computer applications to undergraduates. The criticisms of the early computer-derived classifications may have been justified and even if some scepticism still occurs in phenetics the advantages of automatic identifications, key construction, and mapping cannot be denied. If for no other reason, the storage capacity of a computer enables so much more information to be kept and retrieved that it allows completion of projects which were hitherto too large and complex. Publications are always limiting in format and size. The building up of computer networks permits information to be amassed by direct input from many sources. These assets help to build identikits for such things as medical diagnosis and pest damage as well as identification routines. Recently biologists have been working on identification in a wider context: for example, chromosome modelling and mapping of evolutionary trends based on automatic calculation from stored data. We have been working on similar techniques for graph plotting, contouring and modelling of animal classifications (Wilkinson, 1973a). Colleagues are also trying to determine speciation and hybridization from electrophoresis and chromatography and we are hoping to interpret the results, e.g. range of widths and positions of spectral banding for each taxon, by computer. When visual input units are better developed this will make the task much easier.

To answer further those who have questioned why the taxonomist spends time developing programs for key construction and identification—indeed why we need to use computers at all when research scientists can identify specimens for us—I suggest that at least part of the answer is as follows:

Firstly, many taxonomists are primarily interested in the fundamental research of their groups and therefore too much routine identification becomes a burden especially when they need to be convinced that these labours are worthwhile. To use a research scientist for carrying out this work is an expensive procedure, especially if all that is required is a list of names to fill out an annual report. If accurate computerized keys and identification programs can be produced, computer operators with some knowledge of the biology and nomenclature of the group can be trained to carry out this work, so leaving the expert to his research. Of course, where serious consequences are likely to ensue from determinations, medical diagnosis or criminal proceedings for example, it would then be essential to bring in the expert to confirm electronic identifications.

Secondly, there can be considerable delay in carrying out identifications, not only because specimens may have to be sent long distances by post but also in waiting their turn to be determined. This could have serious consequences if they are urgent, as insect pests might be. Devastation of crops could be complete by the time the species was named and control measures were brought into operation. With a terminal on site, identification and control methods could be ascertained rapidly.

Thirdly, one of the best ways to become conversant with a group would be to serve as a computer operator and whilst carrying out a useful function give oneself computer-assisted instruction.

The increasing use of computers in the biological sciences is already providing more jobs for young graduates. Those students graduating at the bachelor level who do not remain at University and/or train as specialists could be well suited to filling the niche of a computer operator whose function it would be to carry out automatic identifications and the extracting of relevant computer stored information.

Judging from the interest shown in the teaching of computer biology it would seem appropriate to make the following observations about one or two of the current courses, related to systematics.

Recently taxonomy has become respectable again as a subject for teaching in higher education. A number of Universities and Polytechnics have given improved emphasis to this subject in all aspects. Many consider some taxonomic instruction essential to all biology undergraduates. London University have even instituted an "Evolution and Taxonomy" final year specialist subject as an optional equivalent to subjects like Marine Ecology, Parasitology, Entomology etc. for those students reading for external degrees. This is a course of some 150 teaching hours involving methods of identification, principles of classification, rules of nomenclature and the species problem in which one can introduce not only the classical definition and identification methods but also the numerical and phenetic concepts. Our C.N.A.A. (Council for National Academic Awards) degree specifically introduces computer techniques and students carry out automatic identification and description routines as well as key construction, classification programs and related ecological problems. The computer aspects evoke considerable enthusiasm and comparisons are made between numerical and non-numerical classifications of imaginary animals, caminacules (Sokal, 1966) and heraldic animals as well as actual fauna and flora. Hypothetical animals help one to get away from pre-established biological thinking.

In a new interfaculty master's degree where similar statistics and computer techniques are required in different disciplines, students spend much of the first year, of a 2 year course, being taught in the Mathematics and/or Computer Departments. The second year is spent in the student's own department working on a project which involves considerable computational expertise for which taxonomic and ecological problems lend themselves. A bonus of this course lies in the fact that students realize that those from very different disciplines (e.g. Management or Business Studies) may solve their problems in similar ways.

Thus many aspects of biological research, not least those mentioned here, show why there is an increasing need to train biologists with some knowledge of computers. They will be given an awareness of the types of biological problem that can be aided by computer and will be provided with potentially better job prospects. The importance of teaching computer biology is reflected in that two symposia are planned for early 1974—one organized by the Systematics Association on the teaching of systematics and the other at Lancaster University on the teaching of "Computers in Higher Education".

REFERENCES

SOKAL, R. R. (1966). *Scient. Am.* **215,** 106–116.
WILKINSON, C. (1973a). Graphical methods for representing computer classifications. *J. Ent. (B)* **42**(1), 103–112, 7 figs.
WILKINSON, C. (1973b). Computers in Biology. *Can. Ent.* **105,** 1193–1197.

Classified Bibliography
of Computers and Identification

R. J. PANKHURST

The titles given here were selected by the editor, and are intended to cover the most important literature on the subject. Those who wish to study the subject may consult the following index to discover titles of interest. The references are numbered in alphabetical order by first author name. Other references may be found with the papers in this volume, particularly that of Morse (p. 42).

Subject	Reference nos
TECHNIQUES	
Key-construction	12, 13, 32, 34, 35, 48–50, 52, 59–62, 69, 70, 73, 86, 91, 97.
Matching/comparison	7–10, 28–30, 33, 36, 38, 55, 58, 65, 73, 80, 88, 96.
On-line programs	6, 23, 43, 50, 52, 54, 68, 74.
Polyclaves	14, 52, 67.
Probabilistic methods	4, 15, 25, 44, 45, 57, 89, 95.
Tabular keys	46, 56.
Continuous identification and classification	31, 74–79.
Identification with automatic specimen description	22, 72.
Data capture	11, 42, 63.
Character selection	1, 2, 24, 28, 52, 71, 94, 95.
Theory	20, 27, 36a, 37, 40, 41, 47, 82, 82a, 97.
Description of taxa	39, 66, 81.
Data standards	51
Errors	82a
APPLICATIONS	
Bacteria	3, 10, 15, 22, 26, 28–30, 44, 45, 54, 57, 58, 71, 72, 74–79, 83, 94, 95.
Yeasts	1, 2, 9.

Subject	Reference nos
APPLICATIONS	
Fungi	43
Botany, higher plants	4, 19, 70, 80, 91–93, 96.
Pollen	21, 38, 84, 88.
Zoology	56, 85.
Medicine	7, 8, 16–18, 36, 55, 90.
GENERAL	
Surveys and reviews	36a, 53, 64.
Historical	87, 89.
Teaching	5, 19, 26.

References

1. BARNETT, J. A. (1971). Selection of tests for identifying yeasts. *Nature, Lond.* **232**(33), 221–223.
2. BARNETT, J. A. and PANKHURST, R. J. (1974). "A New Key to the Yeasts". North-Holland, Amsterdam.
3. BASCOMB, S., LAPAGE, S. P., CURTIS, M. A. and WILLCOX, W. R. (1973). Identification of bacteria by computer, Identification of reference strains. *J. gen. Microbiol.* **77**, 291–315.
4. BAUM, B. R. and LEFKOVITCH, L. P. (1972). A model for cultivar classification and identification with reference to oats (Avena). II. A probabilistic definition of cultivar groupings and their bayesian identification. *Can. J. Bot.* **50**, 131–138.
5. BEAMAN, J. H. Computers in some instructional aspects of systematic botany. This volume, Chapter 16.
6. BOUGHEY, A. S., BRIDGES, K. W. and IKEDA, A. G. (1968). An automated biological identification key. Mus. Syst. Biol., Univ. Calif., Irvine.
7. BRODMAN, K. (1960). Diagnostic decisions by machine. IRE *TRANS. on medical electronics*, ME-7, **3**, 216–219.
8. BRODMAN, K., VAN WOERKOM, A. J., ERDMANN, A. J. and GOLDSTEIN, L. S. (1959). Interpretation of symptoms with a data-processing machine. *Archs int. Med.* **103**, 116–122.
9. CAMPBELL, I. (1973). Computer identification of the genus *Saccharomyces*. *J. gen. Microbiol.* **77**, 127–135.
10. COWAN, S. T. and STEEL, K. J. (1960). A device for the identification of micro-organisms. *Lancet* **1**, 1172–1173.
11. CROVELLO, T. J. (1968). The effects of missing data and of two sources of character values on a phenetic study of the willows of California. *Madroño* **19**(8).
12. DALLWITZ, M. J. (1974). A flexible computer program for generating diagnostic keys. *Syst. Zool.* **23**(1), 50–57.
13. DALLWITZ, M. J. (1974). User's guide to KEY, a computer program for generating identification keys. CSIRO, Canberra.

14. DUKE, J. A. (1969). On tropical tree seedlings. I. Seeds, seedlings, systems and systematics. *Ann. Mo. bot. Gdn.* **56**(2), 125–161.

15. DYBOWSKI, W. and FRANKLIN, D. A. (1968). Conditional probability and the identification of bacteria. *J. gen. Microbiol.* **54**, 215–229.

16. EDWARDS, D. A. W. (1970). Flowcharts, diagnostic keys, and algorithms in the diagnosis of Dysphagia. *Scot. med. J.* **15**, 378–385.

17. EDWARDS, D. A. W. and PANKHURST, R. J. (1974). The development of a diagnostic key for *Dysphagia*. Proceedings of "Informatique Medicale", IRIA Toulouse.

18. FREEMON, F. R. (1972). Medical diagnosis, comparison of human and computer logic. *Biomed. Comp.* **3**, 217–221.

19. FURLOW, J. J. and BEAMAN, J. H. (1971). Sample Taxonomic data matrices for vascular plants prepared by students of Michigan State University. *Flora N. Am. Rep.* **56**, 1–118.

20. GAREY, M. R. (1970). "Optimal Binary Decision Trees for Diagnostic Identification Problems". PhD Thesis, Wisconsin.

21. GERMERAAD, J. G. and MULLER, J. (1970). A computer-based numerical coding system for the description of pollen grains and spores. *Rev. Palaeobot. Palynol.* **10**, 175–202.

22. GLASER, D. A. and WARD, C. B. (1972). Computer identification of bacteria by colony morphology. *In* "Frontiers in Pattern Recognition" (S. Watanabe, ed.), pp. 139–162. Academic Press, New York and London.

23. GOODALL, D. W. (1968). Identification by computer. *BioScience* **18**(6), 485–488.

24. GOWER, J. C. and BARNETT, J. A. (1971). Selecting tests in diagnostic keys with unknown responses. *Nature, Lond.* **232**(5311), 491–493.

25. GOWER, J. C. Relating classification to identification. This volume (Chapter 15).

26. GRIMES, G. M., RHOADES, H. E., ADAMS, F. C. and SCHMIDT, R. V. (1972). Identification of bacteriological unknowns, a computer-based teaching program. *J. med. Ed.* **47**, 289–292.

27. GUIASU, S. (1970). On an algorithm for recognition. *In* "Mathematics in the Archaeological and Historical Sciences". Proc. Anglo-Romanian Conference (Hudson, ed), 96–102, University of Edinburgh.

28. GYLLENBERG, H. G. (1963). A general method for deriving determination schemes for random collections of microbial isolates. *Ann. Acad. sci. fenn.* A. IV. **69**, 5–23.

29. GYLLENBERG, H. G. (1965). A model for computer identification of micro-organisms. *J. gen. Microbiol.* **39**, 401–405.

30. GYLLENBERG, H. G. and NIEMELÄ, T. K. New approaches to automatic identification of micro-organisms. This volume (Chapter 9).

31. GYLLENBERG, H. G. and NIEMELÄ, T. K. Simulation of computer-aided self-correcting classification method. This volume (Chapter 10).

32. HALL, A. V. (1970). A computer-based system for forming identification keys. *Taxon* **19**(1), 12–18.

33. HALL, A. V. (1969). Group forming and discrimination with homogeneity functions. *In* "Numerical Taxonomy" (A. J. Cole, ed.), pp. 53–68. Academic Press, London and New York.

34. HALL, A. V. (1973). The use of a computer-based system of aids for classification. *Contr. Bolus Herb.* **6**, University of Cape Town.

35. HALL, A. V. A system for automatic key-forming. This volume (Chapter 3).
36. HALL, P., HALLEN, B. and SELANDER, H. (1971). Linear discriminatory analysis, a patient classifying method for research and production control. *Meth. Inf. Med.* **10**(2), 96–102.
36a. HILL, L. R. (1974). Theoretical aspects of numerical identification. *Int. J. Syst. Bact.* **24**(4), 494–499.
37. HILLS, M. (1967). Discrimination and allocation with discrete data. *Appl. Statist.* **16**(3), 237–250.
38. HANSEN, B. S. and CUSHING, E. J. (1973). Identification of pine pollen of late quaternary age from the Chuska Mountains, New Mexico. *Geol. Soc. Am. Bull.* **84**, 1181–1200.
39. JARDINE, N. and SIBSON, R. (1970). Quantitative attributes in taxonomic descriptions. *Taxon* **19**(6), 862–870.
40. JIČÍN, R. (1972). Some problems of description of sets of objects. *J. theor. Biol.* **34**, 295–311.
41. JIČÍN, R., PILOUS, Z. and VAŠÍČEK, Z. (1969). Basis for a formal method for the formation and evaluation of diagnostic keys. (In German) *Preslia* **41**, 71–85.
42. KELLER, C. and CROVELLO, T. J. (1973). Procedures and problems in the incorporation of data from floras into a computerized data bank. *Indiana Acad. Sci.* **82**, 116–122.
43. KENDRICK, B. (1972). Computer graphics in fungal identification. *Can. J. Bot.* **50**, 2171–2175.
44. LAPAGE, S. P., BASCOMB, S., WILLCOX, W. R. and CURTIS, M. A. (1970). Computer identification of bacteria. *In* "Automation, Mechanization and Data Handling in Microbiology" (A. Baillie and R. J. Gilbert, eds), 1–22. Academic Press, London and New York.
45. LAPAGE, S. P., BASCOMB, S., WILLCOX, W. R. and CURTIS, M. A. (1973). Identification of bacteria by computer, general aspects and perspectives. *J. gen. Microbiol.* **77**, 273–290.
46. LEENHOUTS, P. W. (1966). Keys in biology. *Proc. K. ned. Akad. Wet.* **69C**, 571–596.
47. MACDONALD, D. K. C. (1952). Information theory and its application to taxonomy. *J. appl. Phys.* **23**(5), 529–531.
48. METCALF, Z. P. (1954). The construction of keys. *Styst. Zool.* **3**, 38–45.
49. MORSE, L. E., BEAMAN, J. H. and SHETLER, S. G. (1968). A computer system for editing diagnostic keys for Flora North America. *Taxon* **17**(5), 479–483.
50. MORSE, L. E. (1971). Specimen identification and key construction with time-sharing computers. *Taxon* **20**(1), 269–282.
51. MORSE, L. E., PETERS, J. A. and HAMEL, P. B. (1971). A general data format for summarizing taxonomic information. *BioScience* **21**(4), 174–181.
52. MORSE, L. E. (1974). Computer programs for specimen identification, key construction and description printing. *Publs. Mus. Mich. St. Univ., Biol.* **5**(1).
53. MORSE, L. E. Recent advances in the theory and practice of biological specimen identification. This volume (Chapter 2).
54. MULLIN, J. K. (1970). COQAB-A computer optimized question asker for bacteriological specimen identification. *Math. BioSciences* **6**, 55–66.
55. NASH, F. A. (1960). Diagnostic reasoning and the logoscope. *Lancet*, 1442–1446.
56. NEWELL, I. M.(1970). Construction and use of tabular keys. *Pacific Insects* **12**(1), 25–37.

57. NIEMELÄ, S. I., HOPKINS, J. W. and QUADLING, C. (1968). Selecting an economical binary test battery for a set of microbial cultures. *Can. J. Microbiol.* **14**(3), 271–279.

58. OLDS, R. J. (1970). Identification of bacteria with the aid of an improved information sorter. *In* "Automation, Mechanization and Data Handling in Microbiology" (A. Baillie and R. J. Gilbert, eds), pp. 85–89. Academic Press, London and New York.

59. PANKHURST, R. J. (1970). A computer program for generating diagnostic keys. *Computer J.* **12**, 2, 145–151.

60. PANKHURST, R. J. and WALTERS, S. M. (1971). Generation of keys by computer. *In* "Data Processing in Biology and Geology" (J. L. Cutbill, ed.), pp. 189–203. Academic Press, London and New York.

61. PANKHURST, R. J. (1970). Key generation by computer. *Nature, Lond.* **227**, 1269–1270.

62. PANKHURST, R. J. (1971). Botanical keys generated by computer. *Watsonia* **8**, 357–368.

63. PANKHURST, R. J. (1972). A method for data capture. *Taxon* **21**(5/6), 549–558.

64. PANKHURST, R. J. (1974). Automated identification in systematics. *Taxon* **23**(1), 45–51.

65. PANKHURST, R. J. "Identification by matching". This volume (Chapter 6).

66. PANKHURST, R. J. Identification methods and the quality of taxonomic description. This volume (Chapter 14).

67. PANKHURST, R. J. and AITCHISON, R. R. A computer program to construct polyclaves. This volume (Chapter 5).

68. PANKHURST, R. J. and AITCHISON, R. R. An on-line identification program. This volume (Chapter 12).

69. PAYNE, R. W. Genkey: a program for constructing diagnostic keys. This volume (Chapter 4).

70. PETTIGREW, C. J. and WATSON, L. (1973). On identification of sterile Acacias and the feasibility of establishing an automatic key-generating system. *Aust. J. Bot.* **21**, 141–150.

71. PIGUET, J. D. and ROBERGE, P. (1970). Problèmes posés par le diagnostic automatique des bâtonnets gram-négatifs. *Can. J. Publ. Hlth* **61**, 329–335.

72. RIDDLE, J. W., KABLER, P. W., KENNER, B. A., BORDNER, R. H., ROCKWOOD, S. W. and STEPHENSON, H. J. R. (1956). Bacterial identification by infrared spectrophotometry. *J. Bact.* **72**, 593–603.

73. ROSS, G. J. S. Rapid techniques for automatic identification. This volume (Chapter 7).

74. RYPKA, E. W. (1971). Automatic identification, data accumulation, analysis and reclassification. 71st Meeting, Am. Soc. for Microbiology.

75. RYPKA, E. W. and BABB, R. (1970). Automatic construction and use of an identification scheme. *Med. Res. Engng* **9**(2), 9–19.

76. RYPKA, E. W. (1971). Truth table classification and identification. *Space Life Sciences* **3**, 135–156.

77. RYPKA, E. W., CLAPPER, W. E., BOWEN, I. G. and BABB, R. (1967). A model for the identification of bacteria. *J. gen. Microbiol.* **46**, 407–424.

78. RYPKA, E. W. (1972). "Pattern Recognition and Continuous Classification". IEEE Pattern Recognition Workshop.

79. RYPKA, E. W. "Pattern recognition and microbial identification". This volume (Chapter 11).

80. SCHEINVAR, L. and HELIA BRAVO, H. (1974). Ensayo de Utilización de Computadoras

para Identificación de Especies del Género Neobuxbaumia (Cactaceae). (In Spanish, with English summary.) *Anls Inst. biol. Univ. nac. Aut. Mex.* ser Bot. **44**.

81. SHETLER, S. G. A generalized descriptive data bank as a basis for computer-assisted identification. This volume (Chapter 13).

82. SLAGLE, J. R. (1964). An efficient algorithm for finding certain minimum-cost procedures for making binary decisions. *J. ACM.* **11**(3), 253–264.

82a. SNEATH, P. H. A. (1974). Test reproducibility in relation to identification. *Int. J. Syst. Bact.* **24**(4), 508–523.

83. STEEL, K. J. (1962). The practice of bacterial identification. *Symp. Soc. gen. Microbiol.* **12**, 405–432.

84. TAGGART, R. E. (1971). Automation of identification procedures in palynology using taxonomic data matrices. *Flora N. Am. Rep.* **63**, 67–90.

85. TURNER, B. D. (1972). Taxonomic and ecological studies of the Psocoptera of Jamaica, 90–92, from thesis, Leeds.

86. VERHELST, M. (1969). A technique for constructing decision tables. *IAG Q. J.* **2**(1), 27–36.

87. VOSS, E. G. (1952). The history of keys and phylogenetic trees in systematic biology. *J. Sci. Labs Denison Univ.* **43**, 1–25.

88. WALKER, D., MILNE, P., GUPPY, J. and WILLIAMS, J. (1968). The computer assisted storage and retrieval of pollen morphological data. *Pollen et Spores* **10**(2), 251–262.

89. WALTERS, S. M. Traditional methods of biological identification. This volume (Chapter 1).

90. WARNER, H. R., RUTHERFORD, B. D. and HOUTCHENS, B. A. (1972). Sequential bayesian approach to history taking and diagnosis. *Comp. Biomed. Res.* **5**, 256–263.

91. WATSON, L. and MILNE, P. (1972). A flexible system for automatic generation of special purpose dichotomous keys and its application to Australian grass genera. *Aust. J. Bot.* **20**, 331–352.

92. WEBSTER, T. (1969). Developments in the description of potato varieties. Part I—Foliage. *J. natn. Inst. agric. Bot.* **11**(3), 455–475.

93. WEBSTER, T. (1970). Developments in the description of potato varieties. Part II—Inflorescences and tubers. *J. natn. Inst. agric. Bot.* **12**, 17–45.

94. WILLCOX, W. R. and LAPAGE, S. P. (1972). Automatic construction of diagnostic tables. *Comp. J.* **15**(3), 263–267.

95. WILLCOX, W. R. and LAPAGE, S. P. Methods used in a program for computer-aided identification of bacteria. This volume (Chapter 8).

96. WILMOTT, A. J. (1950). A new method for the identification and study of critical groups. *Proc. Linn. Soc.* **162**(1), 83–98.

97. WINSTON, P. (1969). A heuristic program that constructs decision trees. MIT Project MAC. Artificial Intelligence Memo 172.

Bibliography of Computer Programs

A selected bibliography of computer programs for identification is presented. Programs are listed in alphabetical order by name. In each case, the mention of a program implies only that the author is willing to make it available, and not necessarily that he is willing or able to embark on program conversion or development on behalf of others.

Only programs which are directly concerned with identification or data capture are listed. Other related programs exist, such as for description printing, and conversion of data from one format to another, but are not included. The list may still not be representative, for several reasons:

1. Several programs which are being revised at the time of writing are not listed.
2. Programs which are highly machine dependent are omitted. This is particularly the case for data capture.

Potential users of these programs should realize that the information cited may go out of date quite rapidly, and that it is wisest to keep in contact with the author of a program in order to keep abreast with improvements and new developments. Addresses are not given if the author has contributed a paper to this volume: see the appropriate paper.

A brief classification of the programs follows:

Key-forming	CLASP, GENKEY, KEY, KEYS, KEY2, KEY2M3
On-line identification	IDENT4, ONLINE
Punched card keys (polyclave)	CARDKEY, CDKEY, POLPUN
Comparison of taxa	CMPARE, TPB5C
Identification by comparison	IDENT/NEO, MATCH
Identification by elimination	IDENT/NEO
Test selection	TPM7B
Data capture	TUN08

Name	CARDKEY
Description	This program produces punched-card multiple entry keys (polyclaves) from a data matrix in either of two formats.
Author	L. E. Morse.
Language	FORTRAN IV, General Electric Mark 2 dialect.
Documentation	*Publications of the Museum, Michigan State University, Biological Series* **5**, 1, East Lansing, U.S.A. (1974). $3.

Machine dependence	Designed to operate on the GE Mark II computer network.
Availability	Annotated listing published, see documentation, or material at cost from author.

Name	CDKEY
Description	This program manufactures a punched card multiple-entry key (polyclave). The input data is binary-coded with one taxon per card. An additional card is output which cross-references the taxon names to punch positions in the cards of the key.
Author	Leila M. Shultz, Dept. of Biology, UMC 45, Utah State University, Logan, Utah 84322, U.S.A.
Language	FORTRAN.
Documentation	From author.
Machine dependence	Four library subroutines, property of Marketing Systems Corp., on CDC 6400.
Availability	From author.
Comment	This program was used to produce the Random-Access Key to Genera of Colorado Wildflowers, obtainable from Prof. W. Weber, University of Colorado Museum, Boulder, Colorado 80302, U.S.A.

Name	CLASP
Description	CLASP reads mixed mode data in fixed format, free format or sparse format. Computes weighted similarity coefficients, hierarchical cluster analysis, minimum spanning tree, nearest neighbour table, principal coordinate analysis, density table. Relationships between data matrix, group structure, similarity coefficients and principal coordinates. Reduction of data matrix and similarity matrix from units to groups. Identification by nearest neighbour, added principal coordinates and augmented minimum spanning tree. Binary and batch keys. Treatment of

missing values. Storage and retrieval of structures. Labelled and graphical output. 300 units and 200 variables.

Author	G. J. S. Ross.
Language	A.S.A. FORTRAN IV.
Documentation	Program guide and examples, and this volume, Chapter 4.
Machine dependency	Basic input routine requires character code if not EBCDIC. Short integers are used on ICL 4.70 and IBM 370. Backing store is machine dependent. Names assumed to hold 8 characters. Has been converted to run on CDC 7600.
Availability	To be discussed.

Name	CMPARE
Description	This program is used to compare taxa which are described in a data matrix (two possible formats). The number of differences between all pairs of taxa can be counted and listed. Any one particular taxon can be compared with all the rest, and the number of differences counted.
Author	L. E. Morse.
Language	FORTRAN IV, General Electric Mark 2 dialect.
Documentation	*Publications of the Museum, Michigan State University, Biological Series* **5**, 1, East Lansing, U.S.A. (1974). $3.
Machine dependence	None.
Availability	Annotated listing published, see documentation, or material at cost from author.

Name	GENKEY
Description	Genkey is a computer program for constructing diagnostic keys providing several methods of construction and a wide range of types of output. In particular the program enables the user to influence the choice of tests for use in the key, allows tests to be

L

applied in groups and can also construct probabilistic keys for groups of species which cannot be separated with certainty. Further details are given in "Genkey, a program for constructing diagnostic keys" (this volume, Chapter 4).

Author	R. W. Payne.
Language	FORTRAN IV.
Documentation	User manual, available from the author.
Machine dependencies	The program is at present mounted on an ICL 4/70 computer and it is planned to test it on an IBM 370 in the near future. The only major machine dependencies are in the input and output routines. The input routine would need modification in the recognition of directives and numbers in free format. Some of the output routines would require a package for copying and shifting words stored in integer arrays.
Availability	Cards, EBCDIC code.

Name	IDENT/NEO
Description	This program distinguishes between characters which define a particular genus, those which indicate that the specimen belongs to another genus, those which separate species generally and those which are diagnostic for particular species only. Identification is attempted using characters in each of these classes, using both step-by-step elimination and comparison methods, and the program gives a detailed summary of its results. The program was evaluated using data for the genus Neobuxbaumia in the Cactaceae.
Author	Léia A. de Scheinvar, Departamento de Botanicá, Universidad Nacional Autonoma de México, Apartado Postal 70–268, México 20, D.F.
Language	FORTRAN II on Burroughs 5500.
Documentation	*Anales Inst. biol. Univ. nac. Aut. Mexico, ser. Bot.* **44**.
Machine dependency	None.
Availability	Paper tape, cards or magnetic tape.

Name	IDENT4
Description	This program is intended for interactive identification at a terminal of a time-sharing system. The user specifies characters and their values and step-by-step elimination takes place until a single taxon remains. A "variability limit" can be specified which permits identifications to be made in such a way that the specimen does not have to exactly match the taxa in the data matrix. The program can be asked to suggest which characters might best be used next, and to list the possible taxa by name, at each step in the identification.
Author	L. E. Morse.
Language	FORTRAN IV, General Electric Mark 2 dialect.
Documentation	*Publications of the Museum, Michigan State University, Biological Series* **5**, 1, East Lansing, U.S.A. (1974). $3.
Machine dependence	Slight.
Availability	Annotated listing published, see documentation, or material at cost from author.

Name	KEY
Description	Generates identification keys.
Author	M. J. Dallwitz, CSIRO Division of Entomology, P.O. Box 1700, Canberra City, A.C.T. 2601, Australia.
Language	ANSI FORTRAN, except for two machine-dependent subroutines to store and load the n-th character of an array. The program will run without these subroutines, but with reduced capability for formatting the output.
Documentation	(1) User's guide. (2) M. J. Dallwitz (1974). A flexible computer program for generating identification keys. *Systematic Zoology* **23**, 50–57.
Availability	(1) Listing. (2) Punched cards. Requests for cards should be accompanied by a sample card punched as follows: ABCDEFGHIJKLMNOPQRSTUVWXYZ 0123456789+ − */()$=,. | ;

(3) Magnetic tape. 9-track EBCDIC or ASCII, 800bpi.

Name	KEYS
Description	This is a key-constructing program, and forms part of the BOLAID package, which also includes programs for data-input, editing, group-forming, showing group relations and differences, and printing dendrograms. Modal, binary and multistate date are accepted, characters can be weighted, and taxon frequencies specified. The key is printed in numerical form.
Author	A. V. Hall.
Language	FORTRAN IV.
Documentation	A. V. Hall. The use of a computer-based system of aids for classification. *Contributions from the Bolus Herbarium* no. **6,** University of Cape Town, Rondebosch C.P., South Africa, 1973. See this volume, Chapter 3.
Machine dependence	None.
Availability	Program listing published, see Documentation.

Name	KEY2
Description	This program constructs and prints diagnostic keys in the indented style. Character conveniences and taxon frequencies can be specified and taken into account in key construction. If the taxa in the data matrix are not all distinct from one another, an incomplete key can be produced.
Author	L. E. Morse.
Language	FORTRAN IV, General Electric Mark 2 dialect.
Documentation	*Publications of the Museum, Michigan State University, Biological Series* **5,** 1, East Lansing, U.S.A. (1974). $3.
Machine dependence	None.
Availability	Annotated listing published, see documentation, or materials at cost from author.

Name	KEY2M3
Description	This program constructs and prints diagnostic keys. Binary or multistate characters are allowed, and character weights and taxon frequencies can be used. Either of the two popular styles of key can be produced. If the data does not allow all taxa to be separated, a partial key can be produced. There is an option for printing taxon descriptions.
Author	R. J. Pankhurst.
Language	FORTRAN IV.
Documentation	A writeup is available on request. See also: *Watsonia* **8,** 357–368, 1971, *Nature, Lond.* **227,** 1269–1270, 1970, *Computer J.* **12**(2), 145–151, 1970.
Machine dependence	None.
Availability	Punched cards, paper tape, magnetic tape.

Name	MATCH
Description	This program takes a complete description of a specimen and calculates the similarity between it and every reference taxon in a data matrix. A list of taxa with highest resemblance is printed, from which the identification is completed by conventional means. Agreement in respect of certain characters which are prominent in the specimen can be shown if desired, and also the score of any taxon with which the specimen is thought to be related can be printed.
Author	R. J. Pankhurst.
Language	FORTRAN IV.
Documentation	See this volume, Chapter 6. Also program writeup available on request.
Machine dependence	FORTRAN functions AND and OR are required for logical operations. These are simple assembly code routines.
Availability	Punched cards, paper or magnetic tapes.

Name	ONLINE
Description	This program works in an on-line conversational mode, and requires either a separate computer or a time-shared system. The user selects characters of his specimen and types their values, and the computer eliminates taxa which cannot agree. A list of available characters or of possible taxa can be obtained at any time.
Authors	R. J. Pankhurst, R. R. Aitchison.
Language	FORTRAN IV.
Documentation	See Chapter 12, this volume. Also program writeup available on request.
Machine dependence	None.
Availability	Punched cards, paper or magnetic tape.

Name	POLPUN
Description	This program produces a polyclave (multiple-entry key) on punched cards from a list of characters and a taxonomic data matrix. The cards have an interpretable portion so that the character states and sequence numbers can be printed automatically on the cards.
Authors	R. J. Pankhurst, R. R. Aitchison.
Language	FORTRAN IV on IBM 370, CDC 6400.
Documentation	See Chapter 5, this volume. Also available on request are a program writeup and pamphlet, "How to use a polyclave".
Machine dependence	(1) Requires FORTRAN functions AND and OR, for logical operations. These are simple assembly code routines.
	(2) Has to comply with hardware for punching cards, and operating system software for punching binary cards from FORTRAN at execution time.
Availability	Punched cards, paper or magnetic tape.

Name	TPB5C "Paired Comparisons"
Description	This program takes as input the data matrix output from the program, TUN08. If there are certain tests which are to be used only if necessary for separation, they are read into a "BAD" list. Each organism is compared with all remaining organisms to determine which tests will separate each pair. If there are no tests that give a clean separation, then up to four tests are listed that give partial separation. "Essential" tests, that is tests which alone will separate a given pair of organisms, are stored on a separate disk file. All other sets of tests resulting in separation are stored on another disk file. The minimum inclusive set of tests is then determined.
Author	R. G. Babb, Lovelace Foundation, 5200 Gibson Boulevard S.E., Albuquerque, New Mexico 87108.
Language	ALGOL.
Documentation	No.
Machine dependence	As written, the program has been run only on the Burroughs B5700 computer; with only minor modification it should run on any machine with an ALGOL compiler.
Program availability	Listing and/or card-image tape and/or punched cards.

Name	TPM7B "Best-N Tests"
Description	This program takes as input the output from the program "TPB5C". In general, this program will rearrange the data matrix for N-best tests, considering initially the best ordering in terms of single-test separations, and subsequently adding additional tests based upon their separating power *in addition to* the separation already achieved by the N-number of tests being considered. The program can be run to exclude any tests from the best-test set.
	Subgroups may be listed based upon Truth-Table combinations of the best-N tests. This may be done at a number M which is less-than-equal to the N best

tests, since the total number of subgroups increases as 2^N.

Author	R. G. Babb, Lovelace Foundation, 5200 Gibson Boulevard S.E., Albuquerque, New Mexico 87108.
Language	ALGOL.
Documentation	No.
Machine dependence	As written, the program has been run only on the Burroughs B5700 computer; with only minor modification it should run on any machine with an ALGOL compiler.
Program availability	Listing and/or card-image tape and/or punched cards.

Name	TUN08, "UNSCRAMBLE"
Description	This program accepts data describing the characters of taxa derived from literature, which may conflict if derived from different sources, and combines these together to form a taxonomic data matrix. Binary or multi-state or decimal data can all be dealt with.
Author	R. G. Babb, Lovelace Foundation, 5200 Gibson Boulevard S.E., Albuquerque, New Mexico 87108.
Language	ALGOL.
Documentation	No.
Machine dependence	As written, the program has been run only on the Burroughs B5700 computer; with only minor modification it should run on any machine with an ALGOL compiler.
Program availability	Listing and/or card-image tape and/or punched cards.

A Glossary of Computer-assisted Biological Specimen Identification

L. E. MORSE, R. J. PANKHURST, and E. W. RYPKA

Abstract: A glossary of technical terms used in the literature of biological specimen identification, especially with computer methods. It is intended for biologists familiar with taxonomy but not conversant with computers.

INTRODUCTION

This glossary focuses on specialized terms used in the literature of biological specimen identification, particularly computer-based methods. Although some fundamental taxonomic terms are discussed, this glossary is intended for biologists familiar with taxonomy but not familiar with the terminology of computers and data processing. All principal terms used in the present volume are included; the papers in this volume have been a principal source of our definitions and discussion.

Since computer-assisted identification is currently a very active field of study, preparation of a definitive glossary for the field is impossible. The definitions offered here should be taken as our collective impressions and understandings of the usage of various terms, based on our research experiences, the recent literature, and the presentations and discussion at the 1973 Systematics Association meeting on "Biological Identification with Computers". Further discussion of most of these terms will be found elsewhere in this symposium volume. We thank several reviewers for reading and commenting on our manuscript; particularly helpful critiques were provided by Philip Cantino, Craig Greene, Annette Harkins, Micah I. Krichevsky, Reed C. Rollins, Reed Sutherland, Alice F. Tryon, and Carroll E. Wood Jr.

If sufficient interest is shown, a revised and expanded glossary of this nature could be published elsewhere after a few years. Accordingly, we welcome comments and criticisms, particularly suggestions of additional terms, refinements of our definitions, and clarifications of the distinctions between alternative terms. Such comments may be sent to Larry Morse at the Gray Herbarium, Harvard University, Cambridge, Mass. 02138, U.S.A.

L. E. Morse, R. J. Pankhurst, and E. W. Rypka

GLOSSARY

Absolute affinity. An identification criterion met when the characters of a specimen exceed some prespecified threshold of character-state agreement with the characters of one and only one taxon. (Contrast with *Relative affinity*.)

Algorithm. A finite series of logical steps or instructions by which a specific type of problem can be solved. (The concept of an *algorithm* may be clearer if the use of algorithms in information processing is compared with the use of recipes in cooking.)

Alphanumeric. Composed of letters and/or numbers.

Artificial key. An identification key that is not intended to indicate the natural (evolutionary) relationships of the included taxa. (Contrast with *Natural key*.)

Attribute. See *Character*.

Authority file. A data file or look-up table listing acceptable entries for one or more data fields; used to verify the acceptability of entries for these fields in records being added to other data files. For example, a list of acceptable genus names could be used in checking entries to a taxonomic data file.

Auxiliary character. A character not necessary for the identification of some taxon, yet useful for verifying or confirming that identification. (Also called a *Confirmatory character*; contrast with *Diagnostic character*.)

Batch-processing computer system. A computer system in which a program and all its input data are submitted together to the computer, then run without further human intervention. (Contrast with *Interactive computer system*.)

Bayesian method. A method for calculating the probability that a given unknown belongs to a specified taxon, given (1) some characteristics of the unknown, (2) the probability that a randomly chosen member of the specified taxon would show these characteristics of the unknown, and (3) the probability that a randomly chosen unknown would belong to that taxon. *Likelihood* and other probabilistic methods of identification are similar to the Bayesian method, but do not require *a priori* (advance) knowledge of the expected encounter frequencies for the various taxa. In other words, likelihood calculations require the first two kinds of information listed above, but not the probabilities of unknowns belonging to specified taxa.

Binary character. See *Two-state character.*

Boolean expression. A logical relation involving only true/false statements and the logical operators *and*, *or*, and *not*, often expressed in an algebraic form. When evaluated, a Boolean statement is itself either *true* or *false*. For example, if x is *true* and y is *true*, the value of the Boolean statement "x *and* y" is *true*; but if x is *true* and y is *false*, the value of the same statement is *false*.

Bracketed key. A printed dichotomous key in which the contrasting leads of a couplet are presented together, one immediately following the other, without subordinate couplets intervening. The couplets of a bracketed key are now generally differentiated numerically, and actual brackets or braces are rarely used. (Also called *Bracket key*, *Bracketted key*, and *Parallel key*; contrast with *Indented key*.)

Character. A particular feature or other way in which organisms may differ; a particular basis for comparing two organisms, populations, or taxa. The possible alternative expressions of a character are called the *states*, *values*, or *levels* of the character. (See also *Character state*, *Characteristic*, and *Dependent character*.)

The recognition of character states is a difficult and challenging subject, and furthermore the choice of character states can depend on the purpose of the study. Also, different systems can be employed for scaling and coding the information once a set of character states has been chosen. Therefore, a taxon/character data matrix prepared for one purpose may be of little use for another. For example, careful attention to homology is important in taxonomy and especially in phyletic studies, but less so in identification. Indeed, in identification it is important to consider what a casual observer may *think* he is seeing, regardless of what the actual situation is morphologically.

The terms *attribute*, *property*, *feature*, and *test* are often used synonymously with *character*; generally, this loose terminology causes no problems, as long as the meaning intended is clear from context. Sometimes the terms *character* and *character state* are restricted to generalizations about taxa; the terms *attribute* and *attribute state* are then used to refer to properties of individual organisms or specimens. (Under this terminology, identification involves a comparison of the *attributes* of the unknown specimen with the *characters* of various taxa.) The microbiological term *test*, also often used interchangeably with *character*, refers more properly to the procedures or methodologies used in determining the states or values of various characters.

Character coding. The process of concisely and unambiguously labelling or recording character states with a particular system of numbers, letters, or other symbols.

Character cost. (1) The expense (in time, money, etc.) of determining the state of a specified character for some individual. (2) Any other quantitative indication or estimation of the difficulty or inconvenience of determining the state of some character for a particular individual; a character weight. (See also *Character weighting*.)

Character couplet. Any pair of contrasting and contradictory statements about the possible features of a specimen, organism, or population, especially a pair of contrasting leads in a dichotomous key. Note that a character couplet may be a single two-state character, or may include several different characters, each posing two alternatives. Pairs of contrasting Boolean expressions can also be used as character couplets.

Character hierarchy. See *Dependent character*.

Character list. A list of characters and/or character states utilized in a given context, such as a particular taxonomic data matrix. Also known as a *descriptor list*.

Character nesting. See *Dependent character*.

Character set. (1) Any collection of taxonomic characters. (2) The set of characters available for use in a *multiple-entry key* (*polyclave*). (3) In identification by a *comparison method*, the set of characters for which character states must be determined for each unknown.

Character–set method. See *Comparison method*.

Character–set minimization. Selection of an optimum or near-optimum subset of the available characters for a particular use. Usually this subset is the smallest set adequate for identification, but other considerations such as *character cost* can also be involved. (Also called *Character-set optimization* and *Test-set reduction*.)

Character state. One of two or more possible alternative expressions of a character, also known as *values* or *levels* of the character. Characters may be divided into states *qualitatively* or *quantitatively*, as appropriate for the situation. When gathering data, various *character coding* methods can be used to record the character states. (See also *Character*, *Qualitative character*, and *Quantitative character*.)

Character weighting. A numerical or other indication or estimation of different importance values for the various characters used in some application. (The typical "unweighted" case is really an assumption of equal character weights.)

In identification, unlike classification, it is generally agreed that character weighting is important in giving preference to unambiguous, easily observed characters readily studied in the material to be identified. Such character weights are sometimes called *character costs*. Another use of character weighting in identification is in producing identification systems for different purposes, such as separate keys for flowering, fruiting, and vegetative plant material.

Characteristic. Informally, the possession of a character in a particular state or particular combination of states. For example, we might say that a characteristic of the luna moth (*Actias luna*) is its pale green wings; more formally, this species has the character "wing color" in the state "pale green". The terms *character* and *feature* are also sometimes used in this sense.

Comparability of characters. The requirement in information processing that taxa can be described in a uniform manner so as to allow meaningful comparisons to be made between any of the taxa in a data base. (Also called *Consistency of characters*.)

Comparison chart. See *Data matrix*.

Comparison method. Identification through simultaneous comparison of the unknown's states for a prespecified set of characters with the states for these characters for the various taxa being considered as possible identifications. Many different kinds of *similarity coefficients* can be used to make these comparisons. (Also called *Character-set method*, *matching method*, and *simultaneous method*.)

Computer. A mechanical, electrical, or electronic device or machine which can follow stored instructions to perform operations upon information; these operations are usually logical and/or arithmetical. Often the word *computer* is used to mean only an electronic digital computer.

Computer network. An information system in which the computer(s) and data files can be accessible to various remote terminals, enabling people in different places to use the same hardware, programs, and data.

Computer–readable form. See *Machine-readable*.

Confirmatory character. See *Auxiliary character*.

Consistency of characters. See *Comparability of characters*.

Continuous classification and identification. An approach to taxonomy employing a continuing process of alternating stages of classification and identification, with newly identified examples being incorporated immediately into a revised classification.

Conversational computing. See *Interactive computer system*.

Correlation coefficient. A measure of the "relatedness" of variables, usually the product-moment correlation of Karl Pearson, which ranges from $+1$ to -1. A positive correlation means the variables are directly related, and the greater the value of the correlation coefficient, the higher the correlation. A negative value means the variables are inversely related, and the greater the inverse relationship of the variables, the closer the coefficient approaches -1.

Couplet. See *Character couplet*.

Data bank. See *Data base*.

Data base. A collection of data stored in a machine-readable form available for a particular use. Large data bases are sometimes called *data banks*.

Data chart. See *Data matrix*.

Data field. A single kind of information contained in some or all the records of a data file. Data records are typically composed of several data fields. For example, a data record on geographic localities might contain fields for country, state or province, county, and specific locality. These fields could be of different sizes, and be coded in different ways.

Data matrix (pl. *data matrices*). A rectangular table presenting the states of a number of taxa for each of a list of taxonomic characters, especially when these data are presented in a numerically or symbolically coded machine-readable form. (Also called a *comparison chart, data chart, data table,* or *taxonomic data matrix*.)

Data record. A group of *data fields* treated as a unit in data collection or data processing. For example, a data file on the characters of museum specimens might have one data record for each individual specimen examined. (Also called a *logical record.*)

Data structure. A manner of organizing stored data in a computer memory. Frequently used data structures include arrays or matrices, sequential lists, formatted files, and hierarchical patterns, as well as individually labeled locations in core storage. The choice of an appropriate data structure is fundamental to the design of efficient and economical computer programs. (See also *Format.*)

Data table. See *Data matrix.*

Decision tree. A general term for a process of algorithmic selection among alternatives by choosing, for each case, a particular path through a given branching succession of choices, as in a dichotomous identification key. See *Dichotomous key.*

Dependent character. A character which is applicable to an individual only if that individual shows a different character in a particular state. (For example, "Petal color" as a character applies to an organism or taxon only if it has petals present.) Dependent characters can, in turn, have yet other characters dependent on them, forming a *character hierarchy.* This phenomenon is also known as *character nesting*; in microbiology, such characters are termed *linked tests.*

Descriptor list. See *Character list.*

Diagnostic character. A character whose presence in a particular state confirms or helps to confirm the identification of members of a particular taxon. (Also called a *key character, peculiar character,* or *special character*; contrast with *auxiliary character.*)

Diagnostic key. See *Dichotomous key.*

Diagnostic table. A taxonomic data matrix used in identification by comparing the characteristics of an unknown specimen with those tabulated for various taxa in an attempt to find close agreement and hence suggest an identification.

Dichotomous character. See *Two-state character.*

Dichotomous key. An identification key composed of a series of paired choices between the contrasting alternatives (*leads*) of specified *character couplets*, with each such choice leading either to another character couplet or to the name of a taxon. The dichotomous key is a special (and early) application of the general procedure known as the *decision tree*. A dichotomous key is often also called a *diagnostic key*. A *diagnostic key* is a slightly more general term, since in it, *leads* may occur more than two at a time. However, strictly dichotomous keys are widely preferred.

Direct comparison. Identification of an unknown by direct comparison with previously identified individuals, usually museum specimens.

Discrete character. See *Qualitative character*.

Distance. See *Similarity coefficient*.

Edge–punched cards. Rectangular cards with holes punched near the margin. These holes can be clipped with a hand punch to leave a V-shaped notch. If a deck of these cards is aligned and a needle passed through a particular hole on all the cards, and the deck is then shaken gently, cards notched at the selected position will drop out and others will be retained on the needle. (Also called *edge-notched cards*.)

Edge-punched cards can be used as a multiple-entry key (polyclave) by letting each hole represent a diagnostic characteristic. For each taxon, a card is prepared by notching the positions for characteristics which that taxon shows or sometimes shows. The key is used by choosing a characteristic of the unknown, selecting the cards of taxa having that characteristic, and repeating the process on these cards, using additional characters, until an identification is reached. (An alternative approach is to notch the positions of characters *not* shown by a taxon, so the cards of possible identifications will be *retained* on the sorting needle rather than drop off it.)

Editing. See *Text-editing*.

Elimination method. An identification method in which the characteristics of the unknown are used one by one (or a few at a time) in a repeated process of elimination. Sometimes called the *key method* or the *sequential method*, this process is the basis of *dichotomous keys* and *multiple-entry keys*.

Euclidian distance. See *Similarity coefficient*.

Feature. See *Character*.

Field. See *Data field*.

File. In computer science: (1) an organized collection of machine-readable data records contained contiguously on a single memory device; (2) any collection of data, regardless of actual storage, processed by a computer as if the data were physically a single file. (See also *Data base*.)

Format. The layout or arrangement of computer data for input, output, or off-line storage. Typical uses of formats are in designing printed pages of output, specifying the arrangement of data on punched-card decks, and organizing data on magnetic tapes or other storage media. (See also *Data structure*.)

Group. An informal term for a set of related taxa, often from a restricted geographical area, which is the subject of a particular study.

Hardware. The mechanical equipment, circuitry, and other tangible, permanent components of a computer system, in contrast to the programs (*software*) used to direct the processing of data on the computer.

Identification. The process or result of determining to which group or taxon a particular individual, specimen, population, or other object belongs. Since the possibility that the unknown belongs to a previously unrecognized taxon is not routinely considered, identification assumes the acceptance of a previously established classification for the relevant group of taxa. Identification is sometimes called *recognition* or (incorrectly) *classification*.

Identification matrix. See *Data matrix*.

Identification score. In *probabilistic identification*, a calculation of the likelihood that an unknown belongs to a particular taxon, assuming statistical independence of the various characters used. The calculation is done for each taxon by multiplying together, for all characters observed, the probability that a specimen of that taxon would show the character in the same state as observed for the unknown. The higher this product is for a taxon the more likely it is that the unknown belongs to that taxon. (Also see *Likelihood*.)

Implementation. The programming and other preparations necessary for use of an algorithm on a particular computer. One algorithm can have many implementations, using different programming languages, different data structures, different computers, or different peripheral devices.

Inapplicable character. A *dependent character* which poses a meaningless or unanswerable question about the specimen or taxon at hand. For example, the character "Petal color" is *inapplicable* to plants whose flowers lack petals. (Contrast with *"Unknown" state of a character*.)

Indented key. A printed dichotomous key in which each lead of a couplet is immediately followed by all couplets subordinate to it. In such a key, each subordinate couplet is usually indented one unit farther to the right than its predecessor, giving a stepped outline to the left-hand side of the key. (Contrast with *Bracketed key*.)

Interactive computer system. A computer system to which the user can submit additional information during the execution of a program, permitting the development of a dialogue or conversation between user and computer. (Contrast with *Batch-processing computer system*.)

Key. (1) A device for facilitating identification. (2) A printed dichotomous key.

Key character. See *Diagnostic character*.

Key method. See *Elimination method*.

Keypunch. (1) A manually operated keyboard machine used for punching information onto data-processing cards. (2) To prepare punched cards with a keypunch.

Lead (pronounced "Leed"). One of the two or more contrasting statements bearing the same reference number, especially in a diagnostic key. (See *Character couplet, Dichotomous key*.)

Level. See *Character state*.

Likelihood. A probabilistic estimate of the relative support between or among different hypotheses as explanations for a particular situation. In identification by the likelihood method one estimates, for each taxon, the likelihood that the unknown belongs to that taxon instead of to any other taxon being considered. The *identification score* is one method of calculating likelihoods. (Contrast with *Bayesian method*.)

Linked tests. See *Dependent characters*.

Logical record. See *Data record.*

Machine–dependent. Dependent on special features of a particular kind of computer system or particular peripheral equipment, and therefore not easily implemented elsewhere. (Contrast with *Machine-independent.*)

Machine–independent. Using only hardware and software features generally available on most computer systems, and hence capable of being implemented on other computers with only minimal changes. (Contrast with *Machine-dependent.*)

Machine–readable. Ready for input to a computer without further conversion, such as data on punched cards or magnetic tapes.

Matching method. See *Comparison method.*

Monothetic taxon. A taxon for which one may specify a unique combination of diagnostic characters as being both a necessary and a sufficient criterion for an individual's membership in the taxon. (Contrast with *Polythetic taxon.*)

Multi–access key. See *Multiple-entry key.*

Multiple–entry key. A key which allows users to select for themselves the characters they want to use in identifying each unknown, taking their choices in any order from some character set and using as many or as few characters as necessary for each case. (This freedom of choice contrasts with the prespecified character sequence in a dichotomous key, and with the requirement that all the listed characters be used in the character-set method.)

Multiple-entry keys are also known as *polyclaves, multi-access keys, random-access keys,* and *random-entry keys*; the general procedure employed is called the *elimination method.* Popular kinds of multiple-entry keys include *punched-card keys* and *overlay-card keys,* decks of *edge-punched cards,* and computer-based interactive *polyclaves.*

Multi–state character. A character having three or more discrete and equivalent alternative states, either ordered or unordered. Some quantitative characters are better handled as multistate characters than as continuous ones.

Natural key. A key organized primarily as a synopsis of a classification, indicating the supposed natural (evolutionary) relationships of the included taxa. Such a key is generally more difficult to use, and often less accurate, than a well written artificial key to the group. (Contrast with *Artificial key.*)

Network. See *Computer network.*

Numerical taxonomy. An approach to classification based on the numerical comparison of characters, scored consistently for all the groups being classified. The characters are usually given equal weight, and used in large numbers.

Numerical character. See *Quantitative character.*

On–line computing. See *Interactive computer system.*

Optical scanning. In identification, the automatic formation of a description of a specimen by representing a picture of the specimen as machine-readable data, such as coded levels of grayness on a rectangular grid. Optical scanning methods are also widely used in other kinds of automated picture processing, and in conversion of typed or written information to machine-readable form.

Overlay–card key. See *Polyclave.*

Parallel key. See *Bracketed key.*

Partial key. A key which does not distinguish every included taxon from every other. Consequently, some lead(s) will terminate with the names of more than one taxon.

Pattern recognition. In identification, the use of a scanned image or other characteristic machine-readable pattern as a basis for comparison of the specimen with correspondingly produced patterns for various taxa. In the physical sciences and engineering, the term *pattern recognition* is used more broadly, synonymously with *pattern classification,* for the general topic of comparison, classification, and identification of patterns. *Optical scanning* is a common source of information in pattern recognition.

Peculiar character. See *Diagnostic character.*

Peripheral device. A unit of auxiliary computer equipment (hardware) used in input, output, or other aspects of data handling, but not in actual computations. Common peripherals include card readers, disk or drum memories, tape drives, and high-speed printers.

Polyclave. (1) A multiple-entry punched-card or overlay-card key with one card per characteristic, used in a "peek-a-boo" fashion. (2) Any multiple-entry key.

Polythetic taxon. A taxon for which membership is determined by possession of a large number of a specified set of characteristics, without any particular combination of these being a necessary condition for membership. The normal type of taxon adopted in biological taxonomy.

Probabilistic identification. Any identification method in which some measure is given of the probability that the suggested identification of a given unknown is actually correct, as in the *Bayesian* and *likelihood* methods.

Probability matrix. A taxonomic data matrix recording for each character state an estimate of the probability that an individual specimen of a given taxon will show the character in that state. Such a probability matrix is necessary for most probabilistic identification methods.

Problem–oriented system. A computer system designed to meet a specific purpose, rather than for general purposes.

Program. An expression of an algorithm on a particular computer, or in a particular programming language. (A program includes both instructions and data specifications.)

Property. See *Character.*

Punched–card key. A *multiple-entry key* (*polyclave*) made of punched cards, e.g. data-processing cards.

Qualitative character. A character having states which are discontinuous (discrete) and not counts or measurements, regardless of whether these states can be ordered in sequence. Qualitative characters are either *two-state* or *multi-state* characters. (Contrast with *Quantitative character.*)

Quantitative character. A character concerning counts, measurements, ratios, or other numerical values, also called a *numerical character.* The states of a quantitative character can always be ordered in magnitude, but the division of a quantitative character into a small number of states is often arbitrary. (Contrast with *Qualitative character.*)

Random–access key. See *Multiple-entry key.*

Random–entry key. See *Multiple-entry key.*

Recognition. See *Identification*.

Record. See *Data record*.

Relative affinity. A measure of the similarity of the characteristics of a specimen to the corresponding characters of a taxon, used to rank various taxa as possible identifications when no single taxon is a clear choice. (Contrast with *Absolute affinity*.)

Sequential method. See *Elimination method*.

Similarity coefficient. A measure of the association of character states of two specimens, taxa, or other units. Similarity coefficients can be used in identification in determining the *relative affinity* or *absolute affinity* of a specimen to various taxa in the *comparison method*. Conversely, a *dissimilarity coefficient* is a measure of difference.

A typical simple similarity coefficient is the one calculated by dividing the number of similarities (character matches) by the number of characters compared. Various measures of "taxonomic distance", discussed in works on numerical taxonomy, can also be employed in such comparisons. *Euclidean distance* is a dissimilarity measure based on a representation of taxa as points in a multi-dimensional space, and calculating the distances between them with Pythagoras' theorem.

Simultaneous method. See *Comparison method*.

Software. The programs which direct the processing of information in a computer. (Contrast with *Hardware*.)

Special character. See *Diagnostic character*.

Species list. See *Taxon list*.

Specimen. An individual organism or part of an organism (or a number of small organisms) preserved as a unit for scientific study.

Specimen identification. See *Identification*.

State. See *Character state*.

Suggested identification. A taxon suggested as a possible but unconfirmed identification of an unknown, on the basis of the information available.

Supervisor program. A high-level computer program through which a different program (or programs) must be initiated, and to which the other program(s) return after completion.

Taxon (pl. *taxa*). A taxonomic group of any rank; a group of one or more individuals, populations, or subordinate taxa judged sufficiently similar to each other to be treated together formally as a single named evolutionary or informational unit at a particular rank in the taxonomic hierarchy, and sufficiently different from such other groups of the same rank to be treated separately from them.

For example, the white oak (*Quercus alba*), black oak (*Q. velutina*), and pin oak (*Q. palustris*) are three taxa having the rank of species. They all belong to the oak genus (*Quercus*) which is a taxon having the rank of genus. (The three oak species are thus *subordinate* taxa of the oak genus.) Note that "genus" and "species" are not themselves taxa, but indications of the rank of particular taxa.

Taxon list. A list of the names of the taxa being considered in a particular application, such as a particular taxonomic data matrix. Also called *family list, species list*, etc.

Taxon weighting. A numerical or other indication of different importance values or frequency of occurrence values for the various taxa included in a particular study. Taxon weighting is necessary in *Bayesian* identification, and can be used in many other identification methods.

Taxonomic data matrix. See *Data matrix*.

Terminal. An on-line device through which data may be both submitted to and received from a computer system or information network.

Test. See *Character*.

Test–set reduction. See *Character-set minimization*.

Text–editing (computer–based). Software providing commands for making changes to the information in a program or data file, often with the aid of sophisticated on-line text-searching methods. Some editing systems also provide routines for altering the size or structure of a data file during editing.

Time–sharing computer. A computer that can serve more than one user at once, usually by allocating very short slices of time to each user in rapid rotation.

Truth table. An orderly presentation of the various possible combinations of values of a set of logical propositions or other variables, together with an indication of the truth or falsity of a particular Boolean expression or other logical expression when it is evaluated for each combination of the tabulated values.

For example, two true/false characters could be represented in a four-line truth table having entries for TT, TF, FT, and FF. Each line of the truth table gives a combination of characters which is either possible or impossible as a characterization of an individual of a particular taxon. In identification, the name(s) of possible identifications can be listed after each combination of characteristics, and the truth table can then be used manually or through a computer implementation to find suggested identifications corresponding to various combinations of characteristics.

Two–state character. A character having two distinct and mutually exclusive states. (Also called a *binary character* or *dichotomous character*; contrast with *multi-state character* and with *character couplet*.)

Unknown. In identification, the particular individual, specimen, population, or other sample for which an identification is sought.

"Unknown" state of a character. The condition occurring when the state of a character for a particular individual, specimen, population, or taxon is not included in the information available. (Contrast with *Inapplicable character*.)

User. In computer science, the person using a particular computer program or terminal at a specific time, or more generally, anyone who is concerned with applying a computer for their own purposes.

Value. See *Character state*.

Weighting of characters. See *Character weighting*.

Weighting of taxa. See *Taxon weighting*.

Index

The Systematics Association Publications

1. BIBLIOGRAPHY OF KEY WORKS FOR THE IDENTIFICATION OF THE BRITISH FAUNA AND FLORA. 3rd edition (1967)
   Edited by G. J. KERRICH, R. D. MEIKLE and NORMAN TEBBLE
2. THE SPECIES CONCEPT IN PALAEONTOLOGY (1956)
   Edited by P. C. SYLVESTER-BRADLEY, B.Sc., F.G.S.
3. FUNCTION AND TAXONOMIC IMPORTANCE (1959)
   Edited by A. J. CAIN, M.A., D.Phil., F.L.S.
4. TAXONOMY AND GEOGRAPHY (1962)
   Edited by DAVID NICHOLS, M.A., D.Phil.
5. SPECIATION IN THE SEA (1963)
   Edited by J. P. HARDING and NORMAN TEBBLE
6. PHENETIC AND PHYLOGENETIC CLASSIFICATION (1964)
   Edited by V. H. HEYWOOD, Ph.D., D.Sc. and J. McNEILL, B.Sc., Ph.D.
7. ASPECTS OF TETHYAN BIOGEOGRAPHY (1967)
   Edited by C. G. ADAMS and D. V. AGER
8. THE SOIL ECOSYSTEM (1969)
   Edited by J. G. SHEALS
9. ORGANISMS AND CONTINENTS THROUGH TIME (1973)
   Edited by N. F. HUGHES

LONDON. Published by the Association

Systematics Association Special Volumes

1. THE NEW SYSTEMATICS (1940)
   Edited by JULIAN HUXLEY (Reprinted 1971)
2. CHEMOTAXONOMY AND SEROTAXONOMY (1968)*
   Edited by J. G. HAWKES
3. DATA PROCESSING IN BIOLOGY AND GEOLOGY (1971)*
   Edited by J. L. CUTBILL
4. SCANNING ELECTRON MICROSCOPY (1971)*
   Edited by V. H. HEYWOOD
5. TAXONOMY AND ECOLOGY (1973)*
   Edited by V. H. HEYWOOD
6. THE CHANGING FLORA AND FAUNA OF BRITAIN (1974)*
   Edited by D. L. HAWKSWORTH
7. BIOLOGICAL IDENTIFICATION WITH COMPUTERS (1975)*
   Edited by R. J. PANKHURST

*Published by Academic Press for the Systematics Association